Tests begin on page 36

**Math is Easy Series
Book #1**

**Written By:
April Chloe Terrazas**

Part 1
SUCCESS BOOK

**Use the following resources along with this book
for MAXIMIZED practice and awesomeness!**

*<u>Algebra is Easy Part 1</u>

*<u>Algebra is Easy FULL BOOK WEBINAR
STEP-BY STEP EXPLANATIONS!</u>
-videos of Algebra is Easy Part 1 page by page explained
-videos of Algebra is Easy Part 1 SUCCESS BOOK
page by page explained

*<u>Algebra is Easy Part 2 + SUCCESS BOOK + Webinar</u>

Coming in the Math is Easy Series
- Geometry is Easy
- Algebra 2 is Easy
- PreCalculus is Easy
- Elementary School Math is Easy
- Middle School Math is Easy
- The GED is Easy
- The SAT is Easy

All resources (+ package deals!) available at www.Crazy-Brainz.com

Step by step: See each problem in SUCCESS BOOK solved step-by-step in WEBINAR

Algebra Is Easy: Part 1 SUCCESS BOOK (Workbook + Test Book)
April Chloe Terrazas, BS University of Texas at Austin, Mathematics Tutor since 2004.
Copyright © 2015 Crazy Brainz, LLC
ISBN#: 978-1-941775-26-4

Visit us on the web! **www.Crazy-Brainz.com**

All rights reserved. Published and printed in the United States. This book or any portion may not be reproduced or used in a manner whatsoever without the express written permission of the publisher. This book in no way guarantees any specific grade point average.
Cover design, illustrations and text by: April Chloe Terrazas
Algebra is EASY and YOU CAN DO IT

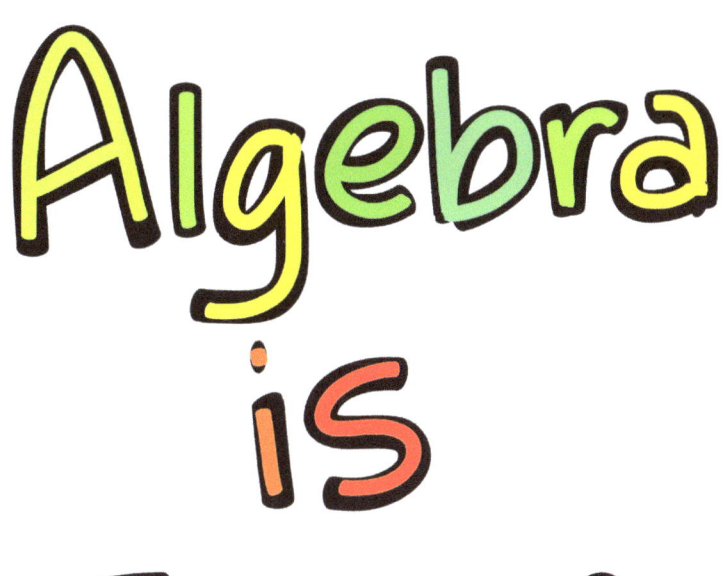

Adding Negative Numbers

Tip: SAME SIGN ADD, DIFFERENT SIGN SUBTRACT.

BE AWESOME! Practice AS MUCH AS POSSIBLE! You deserve THE BEST GRADE POSSIBLE!

Practice:

1. (-4) + (-6) =
2. (-2) + 8 =
3. 22 + (-9) =
4. (-10) + (-40) =
5. 15 + (-15) =
6. 0 + (-7) =
7. (-9) + 15 =
8. (-4) + (-3) =
9. (-1) + 6 =
10. 9 + (-11) =
11. (-25) + (-35) =
12. 12 + (-5) =
13. 0 + (-9) =
14. (-5) + 12 =
15. 8 + (-2) =
16. -4 + (-3) =
17. (-10) + (-10) =
18. 5 + (-5) =
19. 16 + (-4) =
20. -2 + 7 =
21. 7 + (-1) =
22. 120 + (-100) =
23. -5 + (-5) =
24. -3 + 23 =
25. (-20) + 4 =
26. (-18) + (-3) =
27. -16 + (-16) =
28. -16 + 16 =
29. 5 + (-3) =
30. -8 + (-4) =

Practice Answers: (SSA = same side add, DSS = difference sign subtract)

1.	-10	11.	-60	21.	6
2.	6	12.	7	22.	20
3.	13	13.	-9	23.	-10
4.	-50	14.	7	24.	20
5.	0	15.	6	25.	-16
6.	-7	16.	-7	26.	-21
7.	6	17.	-20	27.	-32
8.	-7	18.	0	28.	0
9.	5	19.	12	29.	2
10.	-2	20.	5	30.	-12

You = AMAZING

Subtracting Negative Numbers

Practice:

1. (-3) - (-7) =
2. (-2) - 10 =
3. 1 - (-18) =
4. (-25) - (-40) =
5. 17 - (-17) =
6. 0 - (-9) =
7. (-2) - 13 =
8. (-1) - (-9) =
9. (-2) - 7 =
10. 13 - (-31) =
11. (-12) - (-30) =
12. 15 - (-23) =
13. 0 - (-6) =
14. (-2) - 14 =
15. 6 - (-4) =
16. -3 - 2 =
17. 5 - 9 =
18. (-1) - 7 =
19. 16 - (-3) =
20. (-22) - 12 =
21. (-2) - (-6) =
22. (-1) - 13 =
23. 8 - (-18) =
24. (-5) - (-30) =
25. 17 - (-1) =
26. 0 - (-14) =
27. (-2) - 3 =
28. (-13) - (-9) =
29. (-1) - 6 =
30. 13 - (-11) =

Step by step: See each problem in SUCCESS BOOK solved step-by-step in WEBINAR

Practice Answers: (SSA = same side add, DSS = difference sign subtract)

1.	4	11.	18	21.	4
2.	-12	12.	38	22.	-14
3.	19	13.	6	23.	26
4.	15	14.	-16	24.	25
5.	34	15.	10	25.	18
6.	9	16.	-5	26.	14
7.	-15	17.	-4	27.	-5
8.	8	18.	-8	28.	-4
9.	-9	19.	19	29.	-7
10.	44	20.	-34	30.	24

Multiplying & Dividing #'s & Variables

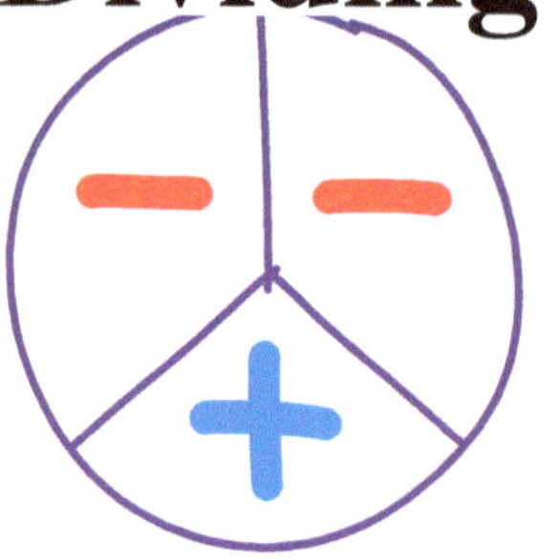

BE AWESOME! Practice AS MUCH AS POSSIBLE! You deserve THE BEST GRADE POSSIBLE!

Practice:

1. (-3)(-7) =
2. (-2) · 10 =
3. 12 • 5 =
4. (-2) x 4 =
5. 11(10) =
6. 16 • 0 =
7. (-2)(-8) =
8. (-1)(-9) =
9. -5 · 7 =
10. 4 x 8 =
11. 3 • 2 • x =
12. (-5)(y) =
13. (20)(2x) =
14. -3 • 7a =
15. (2)(15b) =
16. $20x/5$ =
17. $-15/3$ =
18. $25/(-5)$ =
19. $18/(-3)$ =
20. $-10/-2$ =
21. -100x ÷ -10 =
22. $50/25x$ =
23. $100/-10$ =
24. $16x/-8$ =
25. -18 ÷ 3a =
26. $12x/-2$ =
27. 20xyz ÷ -2 =
28. $22/-11$ =
29. $-28/-4$ =
30. -15ab ÷ 5 =

Practice Answers:

1.	*21*	*6.*	*0*	*11.*	*6x*	*16.*	*4x*	*21.*	*10x*	*26.*	*-6x*
2.	*-20*	*7.*	*16*	*12.*	*-5y*	*17.*	*-5*	*22.*	$2/x$	*27.*	*-10xyz*
3.	*60*	*8.*	*9*	*13.*	*40x*	*18.*	*-5*	*23.*	*-10*	*28.*	*-2*
4.	*-8*	*9.*	*-35*	*14.*	*-21a*	*19.*	*-6*	*24.*	*-2x*	*29.*	*7*
5.	*110*	*10.*	*32*	*15.*	*30b*	*20.*	*5*	*25.*	$-6/a$	*30.*	*-3ab*

You = AMAZING

Order of Operations

Please Excuse My Dear Aunt Sally

PEMDAS

Practice:

1. $4(18 \div 3) =$
2. $-2 + 3 \times 12 =$
3. $8(10 - 7) - 3 =$
4. $56 \div 7 + 4 =$
5. $12 + 3 \times 9 =$
6. $-15 - 4 \div 2 =$
7. $6(12 + 4 \times 2) =$
8. $7 - 2(16 \div 4) =$
9. $2^3 + 9(18 \div 3) =$
10. $35 \div -5 + 4 \times 8 =$
11. $8 + 5^2 - 3(2 \bullet 3) =$
12. $-12 \div 3 + 8(-2) =$
13. $80 \div -8 + 3 \times 4^2 =$
14. $8 \times (-3) + 6 \bullet 6 =$
15. $9 \div 3 + 4 - 12 =$
16. $(-12)^2 - 5 \bullet 8 =$
17. $5(3 \times 2^3 - 5) =$
18. $5(11 + 3) - 2 =$
19. $56 \div 8 - 6 =$
20. $22 + 2 \times 8 =$
21. $6^2 - 4 \times 5 =$
22. $-3(-1 + 7 \bullet 2) =$
23. $5 - 3(24 \div 6) =$
24. $3^3 + 4(24 \div 8) =$
25. $38 \div -2 + 2 \times 6 =$
26. $9 + 6^2 - 2(3 \times 4) =$
27. $-18 \div 3 + 4(-7) =$
28. $60 \div -6 + 2 \bullet 5^2 =$
29. $8 \times (-5) + 6 \bullet 3 =$
30. $12 \div -3 + 5 - 11 =$

Step by step: See each problem in SUCCESS BOOK solved step-by-step in WEBINAR

Practice Answers:

1.	24	6.	-17	11.	15	16.	104	21.	16	26.	21
2.	34	7.	120	12.	-20	17.	95	22.	-39	27.	-34
3.	21	8.	-1	13.	38	18.	68	23.	-7	28.	40
4.	12	9.	62	14.	12	19.	1	24.	-39	29.	-22
5.	39	10.	25	15.	-5	20.	38	25.	-7	30.	-10

Algebra is EASY and YOU CAN DO IT

Distributive Property

Practice:

1. $4(9 + 2) =$
2. $-6(10 - 5) =$
3. $2(12 - 24) =$
4. $5(25 \div 5) =$
5. $-8(7^2 - 25) =$
6. $3(8 * 2 + 13) =$
7. $x(3 + 9) =$
8. $x(x + 4) =$
9. $x(x^2 + 5x - 2) =$
10. $2x(3 + 8) =$
11. $5x(12 \div 4) =$
12. $6x(x + 8) =$
13. $10x(9 + 2) =$
14. $2(7 + 6 - 3^2) + 5^2 =$
15. $7(x + 4) =$
16. $3(5 - 1)^2 =$
17. $-1(x^5y^2 + 8a^2b) =$
18. $4x(3^2 + 7x) =$
19. $12(x + y) =$
20. $4(x^2 + y^2) =$
21. $3a(12 - 7) =$
22. $b(b + 5) =$
23. $b^2(3 + 9 - 8) =$
24. $7a(a + 5) =$
25. $x^2y(x + 4) =$
26. $ab^2(a - b) =$
27. $5x(5x + 5y + 5z) =$
28. $-11(a + b + c)=$
29. $6(x^2 - 7y^3) =$
30. $9(3x - 2y + 6z) =$

Practice AS MUCH AS POSSIBLE! You deserve THE BEST GRADE POSSIBLE!

BE AWESOME!

Practice Answers:

1.	44	11.	$15x$	21.	$15a$
2.	-24	12.	$6x^2 + 48x$	22.	$b^2 + 5b$
3.	-24	13.	$110x$	23.	$4b^2$
4.	25	14.	33	24.	$7a^2 + 35a$
5.	-192	15.	$7x + 28$	25.	$x^3y + 4x^2y$
6.	87	16.	48	26.	$a^2b^2 - ab^3$
7.	$12x$	17.	$-x^5y^2 - 8a^2b$	27.	$25x^2 + 25xy + 25xz$
8.	$x^2 + 4x$	18.	$28x^2 + 36x$	28.	$-11a - 11b - 11c$
9.	$x^3 + 5x^2 - 2x$	19.	$12x + 12y$	29.	$6x^2 - 42y^3$
10.	$22x$	20.	$4x^2 + 4y^2$	30.	$27x - 18y + 54z$

You = AMAZING

Evaluating Algebraic Expressions

Practice:

Evaluate the following expressions for $x = -2$, $y = 4$ and $z = -5$

1. $y \div -2 + x =$
2. $z^2 - y =$
3. $x + y + z =$
4. $8y - 3x =$
5. $2x^3 - 4y^2 =$
6. $8(x - y^3 + z) =$
7. $x(3 + 9y) =$
8. $x(x + 2z) =$
9. $x^2 + 5x - 2 =$
10. $8x + 7z^2 =$
11. $5x(12y + 2) =$
12. $6x(x + 9z) =$
13. $10y(7 + 2y - z) =$
14. $4x^3 + 3y^2 - 8z =$
15. $3z^3(2x^5 + z) =$

Evaluate the following expressions for $a = 3$, $b = -3$ and $c = 4$

16. $a + b - c =$
17. $3a^2 - 6b =$
18. $4a^3 + 5b - 7c =$
19. $12(b + c) =$
20. $4a(b^2 + a^2) =$
21. $3a(12 - 7) =$
22. $b(b + 5) =$
23. $b^2(3a + 9b - 8c) =$
24. $7a(a + 5) =$
25. $a^2b(c + 4) =$
26. $ab^2(a - b) =$
27. $5(5a + 5b + 5c) =$
28. $-11(a + b + c)=$
29. $6(a^2 - 7b^3) =$
30. $9(3a - 2b + 6c) =$

Practice Answers:

1.	*-4*	*11.*	*-500*	*21.*	*45*
2.	*21*	*12.*	*564*	*22.*	*-6*
3.	*-3*	*13.*	*800*	*23.*	*-450*
4.	*38*	*14.*	*56*	*24.*	*168*
5.	*-80*	*15.*	*25, 875*	*25.*	*-216*
6.	*-568*	*16.*	*-4*	*26.*	*162*
7.	*-78*	*17.*	*36*	*27.*	*100*
8.	*24*	*18.*	*65*	*28.*	*-44*
9.	*-8*	*19.*	*12*	*29.*	*1188*
10.	*159*	*20.*	*216*	*30.*	*351*

Step by step: See each problem in SUCCESS BOOK solved step-by-step in WEBINAR

Absolute Value

Practice:

1. $|-3| =$
2. $|2| =$
3. $|-11| =$
4. $-|5| =$
5. $2|-3| =$
6. $-|-21| =$
7. $-3|15| =$
8. $5 + 2|-4| =$
9. $-|-1| - 12 =$
10. $|-120| - |23| =$
11. $-5|-5| =$
12. $3|-12| + 9|-1| =$
13. $-1|-2| - |-8| =$
14. $18 \div |-2| + 3 \bullet 9 =$
15. $-|-30 + 23| =$
16. $-3|12| =$
17. $5|-2| =$
18. $|-125 + 120| =$
19. $7|-12| =$
20. $|26 - 32| =$
21. $|12 + (-20)| + |10| =$
22. $|24 - 14| - |-5| =$
23. $-2|13| =$
24. $7|85 - 76| + 4^2 =$
25. $-4|-4| + (-5 \bullet 3) =$
26. $|-14 \div -2| =$
27. $5^2|-5| =$
28. $-|-3 - 5| - |-12 + 3| =$
29. $|144 \div -12| =$
30. $16 \div |-4| =$

Practice Answers:

1.	*3*	*11.*	*-25*	*21.*	*18*
2.	*2*	*12.*	*45*	*22.*	*5*
3.	*11*	*13.*	*-10*	*23.*	*-26*
4.	*-5*	*14.*	*36*	*24.*	*79*
5.	*6*	*15.*	*-7*	*25.*	*-31*
6.	*-21*	*16.*	*-36*	*26.*	*7*
7.	*-45*	*17.*	*10*	*27.*	*125*
8.	*13*	*18.*	*5*	*28.*	*-17*
9.	*-13*	*19.*	*84*	*29.*	*12*
10.	*97*	*20.*	*6*	*30.*	*4*

Practice AS MUCH AS POSSIBLE! You deserve THE BEST GRADE POSSIBLE!

BE AWESOME!

You = AMAZING

Adding & Subtracting Variables

Practice:

1. $2x + 3x =$
2. $5y + 8y =$
3. $12x - 8x =$
4. $x^2 + 13x^2 =$
5. $-5a + 13a =$
6. $3b + 9b - 26b =$
7. $10x + 4y =$
8. $8ab + 12ab =$
9. $-6r - 13r + 4r =$
10. $18x + 3 =$
11. $4y - 12y =$
12. $100xyz + 50xyz =$
13. $8a + 4a =$
14. $26c - 30c =$
15. $19a + 20b =$
16. $z^2 + 7z^2 =$
17. $-5ab + 8ab - 12ab =$
18. $9x^3 - 13x^3 =$
19. $3y^2 + 4y^3 =$
20. $19x + 21x =$
21. $3a + 5b - 2a + 6b =$
22. $8x^4 + 5y - 3x^4 =$
23. $2d - 8r + 13d =$
24. $5x - 13 + 12x =$
25. $-x + 6x =$
26. $-7y + 8 - 2y =$
27. $12v + 13v =$
28. $-4x + 2(8 + 3x) =$
29. $12 + 4(5y + 2) =$
30. $8b - (12 + 6b) =$

Practice Answers:

1.	$5x$	11.	$-8y$	21.	$a + 11b$
2.	$13y$	12.	$60xyz$	22.	$5x^4 + 5y$
3.	$4x$	13.	$12a$	23.	$15d - 8r$
4.	$14x^2$	14.	$-4c$	24.	$17x - 13$
5.	$8a$	15.	$19a + 20b$	25.	$5x$
6.	$-14b$	16.	$8z^2$	26.	$-9y + 8$
7.	$10x + 4y$	17.	$-9ab$	27.	$25v$
8.	$20ab$	18.	$-4x^3$	28.	$2x + 16$
9.	$-15r$	19.	$3y^2 + 4y^3$	29.	$20y + 20$
10.	$18x + 3$	20.	$40x$	30.	$2b - 12$

Step by step: See each problem in SUCCESS BOOK solved step-by-step in WEBINAR

Multiplying & Dividing Variables

Practice:

1. $6 \bullet 5 \bullet x =$
2. $15 \bullet xy =$
3. $7a \bullet 2a =$
4. $3b^2 \bullet 9b^3 =$
5. $5x(9x^2 + 4x^2) =$
6. $8c \bullet 2c^4 \bullet 5c^3 =$
7. $4y^2(2xy + 3x^2y^4) =$
8. $9z \bullet 4z^2 =$
9. $8a^3(a^4 + 7a^3) =$
10. $7x^2y^4z \bullet 2x^4y^3z^4 =$
11. $3y^4(y^3 + x^2y^2) =$
12. $4a^2 \bullet 3a^4 \bullet 2a^5 =$
13. $3abc(a^2b^3c^4 - a^3b) =$
14. $6x^2y \bullet 9x^3y^4 =$
15. $12xy^4(2x^2y^2 - 3x^4y^2) =$
16. $\frac{5x}{10} =$
17. $20a \div 5 =$
18. $100y^2 \div 2y =$
19. $\frac{7abc}{28a^5} =$
20. $\frac{125x}{25x} =$
21. $(25a^2b^4c^5) \div (50a^3bc^3) =$
22. $\frac{18x^8y^5z^6}{16x^7y^9z^4} =$
23. $\frac{56a^2}{8ab^3} =$
24. $8x^2y \div -4xy^5 =$
25. $\frac{-24x^5y}{12xy^5} =$
26. $\frac{35q^3r^4s^5}{5rs^3} =$
27. $\frac{8}{18xyz} =$
28. $15ab^3 \div 3a^2b=$
29. $\frac{85xy^4z^7}{15x^4yz^9} =$
30. $\frac{16d^3e^6}{4d^8e^3} =$

Practice Answers:

1.	$30x$	11.	$3y^7 + 3x^2y^6$	21.	$b^3c^2/2a$
2.	$15xy$	12.	$24a^{11}$	22.	$9xz^2/8y^4$
3.	$14a^2$	13.	$3a^3b^4c^5 - 3a^4b^2$	23.	$7a/b^3$
4.	$27b^5$	14.	$54x^5y^5$	24.	$-2x/y^4$
5.	$65x^2$	15.	$24x^3y^6 - 36x^5y^6$	25.	$-2x^4/y^4$
6.	$80c^8$	16.	$x/2$	26.	$7q^3r^3s^2$
7.	$8xy^3 + 12x^2y^6$	17.	$4a$	27.	$4/9xzy$
8.	$36z^3$	18.	$50y$	28.	$5b^2/a$
9.	$8a^7 + 56a^6$	19.	$bc/4a^4$	29.	$17y^3/3x^3z^2$
10.	$14x^6y^7z^5$	20.	5	30.	$4e^3/d^5$

You deserve THE BEST GRADE POSSIBLE!

Practice AS MUCH AS POSSIBLE!

BE AWESOME!

You = AMAZING

Practice:

1. $\frac{16}{20}$ =

2. $\frac{2}{10}$ =

3. $\frac{8}{12}$ =

4. $\frac{9}{27}$ =

5. $\frac{4}{26}$ =

6. $\frac{6}{36}$ =

7. $\frac{-3}{9}$ =

8. $\frac{-5}{35}$ =

9. $\frac{16}{20}$ =

10. $\frac{4}{8}$ =

11. $\frac{12}{38}$ =

12. $\frac{11}{77}$ =

13. $\frac{6}{22}$ =

14. $\frac{-2}{20}$ =

15. $\frac{3}{24}$ =

16. $\frac{8}{8}$ =

17. $\frac{4}{12}$ =

18. $\frac{18}{20}$ =

19. $\frac{-5}{15}$ =

20. $\frac{20}{100}$ =

21. $\frac{18}{30}$ =

22. $\frac{49}{56}$ =

23. $\frac{16}{24}$ =

24. $\frac{9}{81}$ =

25. $\frac{13}{26}$ =

26. $\frac{24}{72}$ =

27. $\frac{3}{18}$ =

28. $\frac{-6}{16}$ =

29. $\frac{120}{1000}$ =

30. $\frac{32}{38}$ =

Step by step: See each problem in SUCCESS BOOK solved step-by-step in WEBINAR

Practice Answers:

1.	*4/5*	*6.*	*1/6*	*11.*	*6/19*	*16.*	*1*	*21.*	*3/5*
2.	*1/5*	*7.*	*-1/3*	*12.*	*1/7*	*17.*	*1/3*	*22.*	*7/8*
3.	*2/3*	*8.*	*-1/7*	*13.*	*3/11*	*18.*	*9/10*	*23.*	*2/3*
4.	*1/3*	*9.*	*4/5*	*14.*	*-1/10*	*19.*	*-1/3*	*24.*	*1/9*
5.	*2/13*	*10.*	*1/2*	*15.*	*1/8*	*20.*	*1/5*	*25.*	*1/2*

26.	*1/3*
27.	*1/6*
28.	*-3/8*
29.	*3/25*
30.	*16/19*

Fractions: Improper vs. Mixed

Practice: Convert improper fractions to mixed numbers and mixed numbers to improper fractions

1. $1\frac{3}{7} =$
2. $4\frac{4}{9} =$
3. $2\frac{6}{7} =$
4. $8\frac{3}{4} =$
5. $3\frac{6}{11} =$
6. $5\frac{3}{5} =$
7. $9\frac{4}{13} =$
8. $11\frac{12}{17} =$
9. $13\frac{2}{7} =$
10. $15\frac{3}{7} =$
11. $\frac{12}{5} =$
12. $\frac{40}{7} =$
13. $\frac{13}{3} =$
14. $\frac{24}{7} =$
15. $\frac{28}{4} =$
16. $\frac{23}{5} =$
17. $\frac{36}{9} =$
18. $\frac{11}{3} =$
19. $\frac{56}{7} =$
20. $\frac{41}{5} =$
21. $3\frac{7}{8} =$
22. $\frac{29}{6} =$
23. $5\frac{7}{8} =$
24. $\frac{48}{20} =$
25. $\frac{52}{11} =$
26. $8\frac{6}{11} =$
27. $5\frac{4}{5} =$
28. $\frac{23}{10} =$
29. $\frac{14}{3} =$
30. $9\frac{9}{10} =$

Practice AS MUCH AS POSSIBLE! You deserve THE BEST GRADE POSSIBLE!

BE AWESOME!

Practice Answers:

1.	*10/7*	*6.*	*28/5*	*11.*	*2 2/5*	*16.*	*4 3/5*	*21.*	*31/8*	*26.*	*94/11*
2.	*40/9*	*7.*	*121/13*	*12.*	*5 5/7*	*17.*	*4*	*22.*	*4 5/6*	*27.*	*29/5*
3.	*20/7*	*8.*	*199/17*	*13.*	*4 1/3*	*18.*	*3 2/3*	*23.*	*47/8*	*28.*	*2 3/10*
4.	*35/4*	*9.*	*93/7*	*14.*	*3 3/7*	*19.*	*8*	*24.*	*2 2/5*	*29.*	*4 2/3*
5.	*39/11*	*10.*	*108/7*	*15.*	*7*	*20.*	*8 1/5*	*25.*	*4 8/11*	*30.*	*99/10*

You = AMAZING

Fractions: Multiplying

Practice:

1. $\frac{1}{8} \times \frac{4}{9} =$

2. $\frac{5}{7} \times \frac{14}{15} =$

3. $\frac{2}{3} \times \frac{12}{18} =$

4. $\frac{6}{7} \times \frac{4}{3} =$

5. $\frac{1}{9} \times \frac{3}{4} =$

6. $\frac{5}{6} \times \frac{2}{20} =$

7. $\frac{12}{17} \times \frac{2}{3} =$

8. $\frac{5}{9} \times \frac{6}{35} =$

9. $1\frac{4}{5} \times \frac{5}{9} =$

10. $3\frac{1}{6} \times \frac{8}{9} =$

11. $5\frac{3}{7} \times \frac{4}{9} =$

12. $2\frac{2}{3} \times 1\frac{1}{6} =$

13. $7\frac{1}{2} \times \frac{3}{4} =$

14. $1\frac{3}{7} \times \frac{8}{9} =$

15. $6\frac{2}{5} \times \frac{3}{8} =$

Step by step: See each problem in SUCCESS BOOK solved step-by-step in WEBINAR

Practice Answers:

1.	*1/18*	*6.*	*1/12*	*11.*	*152/63*
2.	*2/3*	*7.*	*8/17*	*12.*	*28/9*
3.	*4/9*	*8.*	*2/21*	*13.*	*45/8*
4.	*8/7*	*9.*	*1*	*14.*	*80/63*
5.	*1/12*	*10.*	*76/27*	*15.*	*12/5*

Practice:

1. $\frac{2}{9} \div \frac{4}{7} =$

2. $\frac{5}{6} \div \frac{2}{18} =$

3. $\frac{14}{15} \div \frac{7}{5} =$

4. $\frac{5}{8} \div \frac{20}{21} =$

5. $\frac{7}{9} \div \frac{3}{7} =$

6. $\frac{6}{11} \div \frac{2}{3} =$

7. $\frac{1}{4} \div \frac{10}{11} =$

8. $\frac{1}{8} \div \frac{3}{16} =$

9. $3\frac{3}{4} \div \frac{4}{9} =$

10. $1\frac{1}{4} \div \frac{8}{9} =$

11. $5\frac{3}{7} \div \frac{6}{21} =$

12. $4\frac{2}{7} \div 2\frac{5}{6} =$

13. $2\frac{7}{8} \div \frac{3}{16} =$

14. $4\frac{3}{4} \div \frac{6}{7} =$

15. $8\frac{1}{3} \div \frac{1}{6} =$

Practice Answers:

1.	*7/18*	*6.*	*9/11*	*11.*	*19*
2.	*15/2*	*7.*	*11/40*	*12.*	*180/119*
3.	*2/3*	*8.*	*2/3*	*13.*	*46/3*
4.	*21/32*	*9.*	*135/16*	*14.*	*133/24*
5.	*49/27*	*10.*	*45/32*	*15.*	*50*

Exponents: Intro

Practice:

1. $4^3 =$
2. $(-3)^4 =$
3. $-(-2)^3 =$
4. $2^5 =$
5. $(-4)^3 =$
6. $3^3 \bullet 4^2 =$
7. $(3 + 2)^2 =$
8. $(18 \div 3)^3 =$
9. $2 \bullet 5^2 =$
10. $-(-2)^4 =$
11. $3^3 \bullet 2^3 =$
12. $(4 - 9)^2 =$
13. $(-x)^3 =$
14. $-(a)^4 =$
15. $(2b)^3 \bullet c^2 =$

Write the following in scientific notation:

16. 4,500 =
17. 204 =
18. 9,120,000 =
19. 1,000,000,000 =
20. 45,600 =
21. 31,200,000 =
22. 7,890,000,000 =
23. 0.012 =
24. 0.00384=
25. 0.716 =
26. 0.0000268 =
27. 0.0021 =
28. 0.091268 =
29. 0.0008 =
30. 0.1034 =

Practice Answers:

1.	64	11.	216	21.	3.12×10^7
2.	81	12.	25	22.	7.89×10^9
3.	8	13.	$-x^3$	23.	1.2×10^{-2}
4.	32	14.	$-a^4$	24.	3.84×10^{-3}
5.	-64	15.	$8b^3c^2$	25.	7.16×10^{-1}
6.	432	16.	4.5×10^3	26.	2.68×10^{-5}
7.	25	17.	2.04×10^2	27.	2.1×10^{-3}
8.	216	18.	9.12×10^6	28.	9.1268×10^{-2}
9.	50	19.	1.0×10^9	29.	8.0×10^{-4}
10.	-16	20.	4.56×10^4	30.	1.034×10^{-1}

Exponents: Basic Uses

Practice: Simplify, no negative exponents.

1. $4^3 \bullet 4^2 =$
2. $2^2 \bullet 2^3 \bullet 2^4 =$
3. $7^8 \bullet 7^{-5} =$
4. $(5a^2)^3 =$
5. $a^5/a^2 =$
6. $6x^3/3x^2 =$
7. $12c^4/4c^{-5} =$
8. $1/y^{-4} =$
9. $(4a^3b^4)^2 =$
10. $(5x^{-5}y^3z^4)^{-2} =$
11. $(2c^5)^3 =$
12. $(5x^2)^{-2} =$
13. $(4xy^2/2x^2y^4)^{-1} =$
14. $(3a \bullet 6b \bullet 4c)^2 =$
15. $(-5y)^{-3} =$
16. $6^2 \bullet 6^2 =$
17. $5^2 \bullet 5^3 =$
18. $10^3 \bullet 10^{-5} =$
19. $a^{-4} =$
20. $xy^{-2} =$
21. $(a^3b^{-2})^{-3} =$
22. $(3x^5)^{-2} =$
23. $(-2ab^{-1})^{-1} =$
24. $(6x^2y^3z^7)^2 =$
25. $8x^2 \bullet (4x^3)^2 =$
26. $7y^{-4} =$
27. $18x^2y^3z^7 \div 9xy^5z =$
28. $25a^{12} \div 50a^8 =$
29. $(81)^{-1} =$
30. $(8a^2b^2c^2)^2 =$

Practice Answers:

1.	1024	6.	$2x$	11.	$8c^{15}$	16.	1296	21.	b^6/a^9	26.	$7/y^4$
2.	512	7.	$3c^9$	12.	$1/25x^4$	17.	3125	22.	$1/9x^{10}$	27.	$2xz^6/y^2$
3.	343	8.	y^4	13.	$xy^2/2$	18.	$1/100$	23.	$b/-2a$	28.	$a^4/2$
4.	$125a^6$	9.	$16a^6b^8$	14.	$5184a^2b^2c^2$	19.	$1/a^4$	24.	$36x^4y^6z^{14}$	29.	$1/81$
5.	a^3	10.	$x^{10}/25y^6z^8$	15.	$1/-125y^3$	20.	x/y^2	25.	$128x^8$	30.	$64a^4b^4c^4$

BE AWESOME! Practice AS MUCH AS POSSIBLE! You deserve THE BEST GRADE POSSIBLE!

Equations: Adding, Subtracting

Practice:

1. $x + 5 = 10$
2. $8 + y = 11$
3. $a + 8 = 21$
4. $5x + 3 = 4x + 7$
5. $b - 4 = 10$
6. $7 = c - 5$
7. $50 + r = 18$
8. $10y + 5 = 11y + 8$
9. $2x - 2 = 3x + 3$
10. $x = 9 + 12$
11. $y - 8 = 11$
12. $3 + c = 25$
13. $17 + r = 23 - 5$
14. $18x + 4 = 17x + 22$
15. $y - 1 = 48$
16. $9 = d + 10$
17. $3 + x = 25$
18. $31 = 3 + k$
19. $24 + m = 33$
20. $16 + y = 25 - 6$
21. $3 + 8x = 7x - 1$
22. $9 + d = 22$
23. $17 = a - 8$
24. $22 = y + 14$
25. $7 + x = 2x$
26. $45 + b = 18$
27. $13 = x - 7$
28. $81 = y + 80$
29. $16 + c = 24$
30. $10x = 9x + 4$

Practice Answers:

1.	$x = 5$	11.	$y = 19$	21.	$x = -4$
2.	$y = 3$	12.	$c = 22$	22.	$d = 13$
3.	$a = 13$	13.	$r = 1$	23.	$a = 25$
4.	$x = 4$	14.	$x = 18$	24.	$y = 8$
5.	$b = 14$	15.	$y = 49$	25.	$x = 7$
6.	$c = 12$	16.	$d = -1$	26.	$b = -27$
7.	$r = -32$	17.	$x = 22$	27.	$x = 20$
8.	$y = -3$	18.	$k = 28$	28.	$y = 1$
9.	$x = -5$	19.	$m = 9$	29.	$c = 8$
10.	$x = 21$	20.	$y = 3$	30.	$x = 4$

Step by step: See each problem in SUCCESS BOOK solved step-by-step in WEBINAR

Equations: Multiplying, Dividing

Practice: simplify as much as possible.

1. $5x = 25$
2. $8x = 88$
3. $28 = -7y$
4. $90 = 10a$
5. $6c = 54$
6. $-4x = 60$
7. $25z = 125$
8. $18x = 36$
9. $2y = 16$
10. $12a = 48$
11. $19x = 38$
12. $3y = 93$
13. $26 = -2p$
14. $32 = 8a$
15. $-11x = 121$
16. $\frac{x}{2} = 5$
17. $12 = \frac{y}{4}$
18. $\frac{x}{-3} = -15$
19. $24 = \frac{a}{2}$
20. $\frac{z}{-5} = -100$
21. $6 = \frac{b}{2}$
22. $\frac{a}{11} = -7$
23. $17 = \frac{y}{3}$
24. $\frac{x}{-8} = 8$
25. $14 = \frac{c}{2}$
26. $\frac{a}{2} = -9$
27. $-5 = \frac{x}{3}$
28. $\frac{y}{7} = 13$
29. $25 = \frac{b}{-2}$
30. $\frac{x}{9} = 12$

Practice Answers:

1.	$x = 5$	11.	$x = 2$	21.	$b = 12$
2.	$x = 11$	12.	$y = 31$	22.	$a = -77$
3.	$y = -4$	13.	$p = -13$	23.	$y = 51$
4.	$a = 9$	14.	$a = 4$	24.	$x = -64$
5.	$c = 9$	15.	$x = -11$	25.	$c = 28$
6.	$x = -15$	16.	$x = 10$	26.	$a = -18$
7.	$z = 5$	17.	$y = 48$	27.	$x = -15$
8.	$x = 2$	18.	$x = 45$	28.	$y = 91$
9.	$y = 8$	19.	$a = 48$	29.	$b = -50$
10.	$a = 4$	20.	$z = 500$	30.	$x = 108$

Equations: Multi-Step

Practice: leave answer in simplest form.

1. $5x + 1 = 36$
2. $7x - 4 = 24$
3. $11x + 18 = 29$
4. $4x - 3 = 13$
5. $8 - 3x = 17$
6. $12x + 6 = 5x + 8$
7. $14 = -2(-3x + 5)$
8. $5(4x - 2) = 12 - 14x$
9. $8x - 3 = 29$
10. $3x - 1 + 5x = 15$
11. $18 = -(x - 4)$
12. $-2x + 16 = 30$
13. $3x - (-15) = 15$
14. $5x + 12 - 3x - 21 = 25$
15. $-6(7x + 14 - 5x) = 12$
16. $x/4 + 9 = 12$
17. $x/3 - 2 = 23$
18. $x/6 + 15 = 25$
19. $9x - 12 = 3x + 15$
20. $-2x + 11 = 17$
21. $3(4x - 1) = 24$
22. $-8x - 8 = 8$
23. $5 - 2(3x - 1) = 12 + 4x$
24. $7x + 13 = 27$
25. $2x - 13 = 5x + 20$
26. $-4(x + 3) = 2(3x - 6)$
27. $8(5x - 3) = 24$
28. $3(2x - 6) = 9$
29. $-2(x + 1) = 10$
30. $25 = 3x + 2 - 5x + 12$

Step by step: See each problem in SUCCESS BOOK solved step-by-step in WEBINAR

Practice Answers:

1.	$x = 7$	11.	$x = -14$	21.	$x = 9/4$
2.	$x = 4$	12.	$x = -7$	22.	$x = -2$
3.	$x = 1$	13.	$x = 0$	23.	$x = -1/2$
4.	$x = 4$	14.	$x = 17$	24.	$x = 40/7$
5.	$x = -4$	15.	$x = -8$	25.	$x = -11$
6.	$x = 2/7$	16.	$x = 12$	26.	$x = 0$
7.	$x = 4$	17.	$x = 75$	27.	$x = 6/5$
8.	$x = 11/17$	18.	$x = 60$	28.	$x = 9/2$
9.	$x = 4$	19.	$x = 9/2$	29.	$x = -6$
10.	$x = 2$	20.	$x = -3$	30.	$x = -11/2$

Absolute Value Equations

Practice: leave answer in simplest form.

1. $|x| = 2$
2. $|x| = 5$
3. $|3x| = 12$
4. $|4x| = 16$
5. $|x + 1| = 2$
6. $|x - 3| = 5$
7. $|x + 9| = 19$
8. $2|5x| = 12$
9. $3|2x| = 21$
10. $4|x - 9| = 36$
11. $5|3x + 3| = 50$
12. $8|2x - 5| = 64$
13. $3|x + 2| = 24$
14. $2|x - 1| + 4 = 12$
15. $7|x + 3| - 2 = 14 - 2$
16. $|x| = 9$
17. $|x| = 120$
18. $|5x| = 15$
19. $|9x| = 45$
20. $|x - 8| = 13$
21. $|x + 5| = 23$
22. $|x - 12| = 54$
23. $3|3x| = 18$
24. $2|8x| = 24$
25. $-6|x + 1| = -42$
26. $7|4x - 3| = 63$
27. $12|3x - 8| = 36$
28. $-11 + 4|x - 2| = 25$
29. $5|2x + 4| - 4 = 16$
30. $9 + 2|8x - 3| = 29$

Practice Answers:

1.	$x = 2, -2$	11.	$x = 7/3, -13/3$	21.	$x = 18, -28$
2.	$x = 5, -5$	12.	$x = 13/2, -3/2$	22.	$x = 66, -42$
3.	$x = 4, -4$	13.	$x = 6, -10$	23.	$x = 2, -2$
4.	$x = 4, -4$	14.	$x = 5, -3$	24.	$x = 3/2, -3/2$
5.	$x = 1, -3$	15.	$x = -1, -5$	25.	$x = 6, -8$
6.	$x = 8, -2$	16.	$x = 9, -9$	26.	$x = 3, -3/2$
7.	$x = 10, -28$	17.	$x = 120, -120$	27.	$x = 11/3, 5/3$
8.	$x = 6/5, -6/5$	18.	$x = 3, -3$	28.	$x = 11, -7$
9.	$x = 7/2, -7/2$	19.	$x = 5, -5$	29.	$x = 0, -2$
10.	$x = 0, 18$	20.	$x = 21, -5$	30.	$x = 13/8, -7/8$

BE AWESOME! Practice AS MUCH AS POSSIBLE! You deserve THE BEST GRADE POSSIBLE!

You = AMAZING

You =

Awesome

Step by step: See each problem in SUCCESS BOOK solved step-by-step in WEBINAR

Graphing on Coordinate Grid (x, y)

Label the following on a coordinate grid.

1. A (4, 0)
2. B (-1, 3)
3. C (6, -3)
4. D (0, -2)
5. E (5, 4)
6. F (-1, 6)
7. G (4, 4)
8. H (0, 3)
9. I (-6, 3)
10. J (2, -4)
11. K (1, 0)
12. L (-5, 7)
13. M (-3, -2)
14. N (-3, 3)
15. O (6, 5)

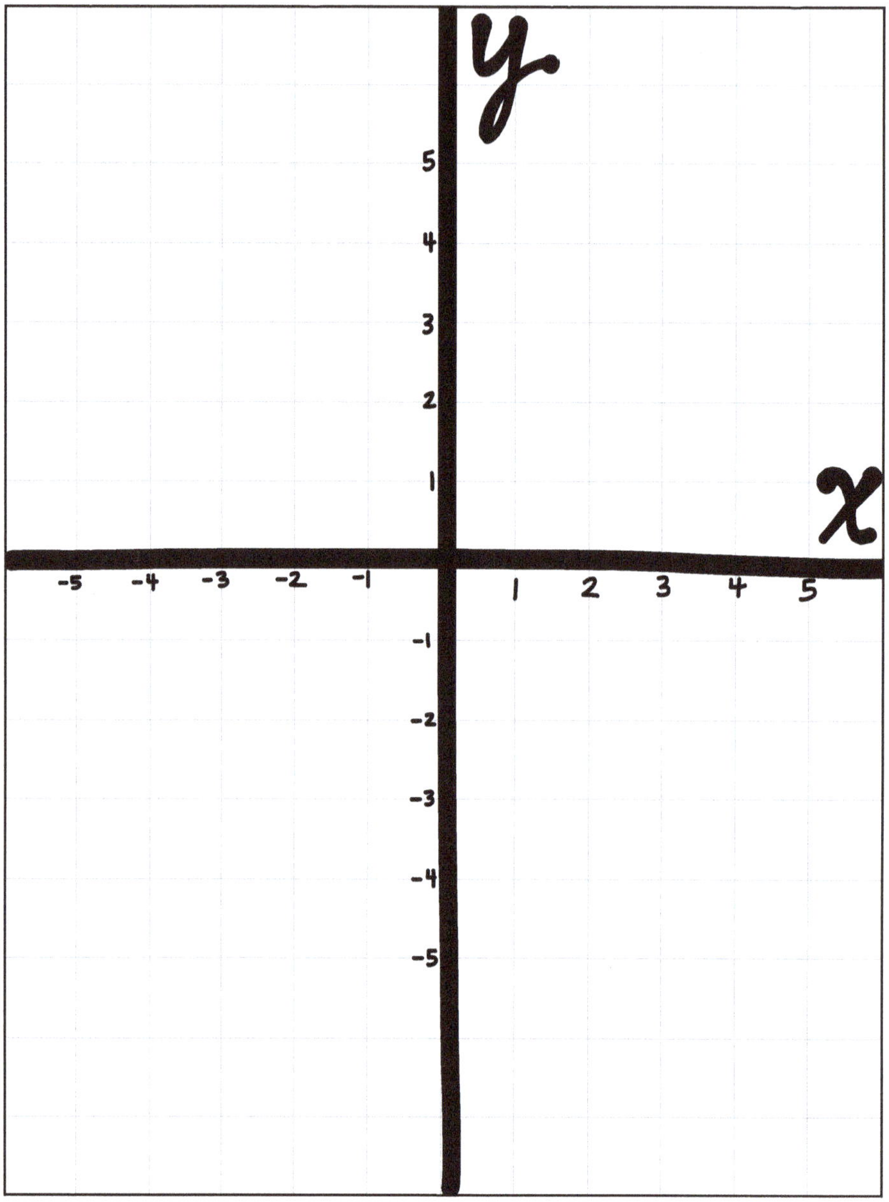

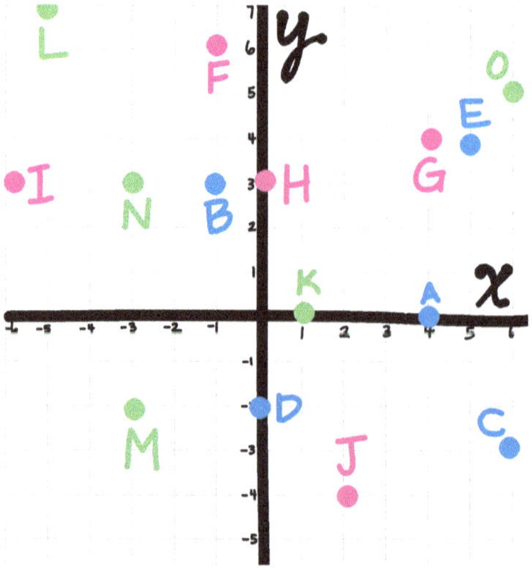

Practice AS MUCH AS POSSIBLE! **You deserve** THE BEST GRADE POSSIBLE!

BE AWESOME!

You = AMAZING

Graphing a Linear Equation Practice

<u>Graph 1 - 15 using a table</u>

1. $y = 2x + 3$
2. $y = 4x - 1$
3. $y = x - 5$
4. $y = \frac{1}{4}x + 3$
5. $y = \frac{2}{3}x - 2$
6. $y = \frac{4}{5}x + 1$
7. $y = -3x + 5$
8. $2x + 3y = 18$
9. $-10x + 5y = 30$
10. $3x - 4y = 12$
11. $y = -x - 4$
12. $y = \frac{2}{5}x + 2$
13. $x + 6y = -24$
14. $y = 4x - 3$
15. $y = -3x + 4$

<u>Graph 16 - 30 using slope intercept form</u>

16. $y = 3x - 1$
17. $y = -2x + 6$
18. $y = 5x - 4$
19. $y = 4x - 2$
20. $y = \frac{1}{3}x + 2$
21. $y = \frac{4}{3}x + 1$
22. $y = \frac{3}{5}x - 5$
23. $y = -x - 1$
24. $y = -4x + 5$
25. $y = \frac{1}{2}x + 6$
26. $y = -3x + 4$
27. $y = 2x - 5$
28. $y = \frac{-2}{5}x - 4$
29. $y = 3x + 3$
30. $y = x + 7$

Answers 1-15 Tables (see reverse side for graphs; you choose your own values for x in the table, these are the points I chose)

1. x	*-1*	*0*	*1*
y	*1*	*3*	*5*
2. x	*-1*	*0*	*1*
y	*-5*	*-1*	*3*
3. x	*-1*	*0*	*1*
y	*-6*	*-5*	*-4*
4. x	*-4*	*0*	*4*
y	*2*	*3*	*4*
5. x	*-3*	*0*	*3*
y	*-4*	*-2*	*0*
6. x	*-5*	*0*	*5*
y	*-3*	*1*	*5*
7. x	*0*	*1*	*2*
y	*5*	*2*	*-1*
8. x	*0*	*3*	*6*
y	*6*	*4*	*2*
9. x	*-2*	*-1*	*0*
y	*2*	*4*	*6*
10. x	*-4*	*0*	*4*
y	*-6*	*-3*	*0*
11. x	*-1*	*0*	*1*
y	*-3*	*-4*	*-5*
12. x	*-5*	*0*	*5*
y	*0*	*2*	*4*
13. x	*-3*	*0*	*3*
y	*-3.5*	*-4*	*-4.5*
14. x	*-1*	*0*	*1*
y	*-7*	*-3*	*1*
15. x	*-1*	*0*	*1*
y	*7*	*4*	*1*

Graphing a Linear Equation Answers:

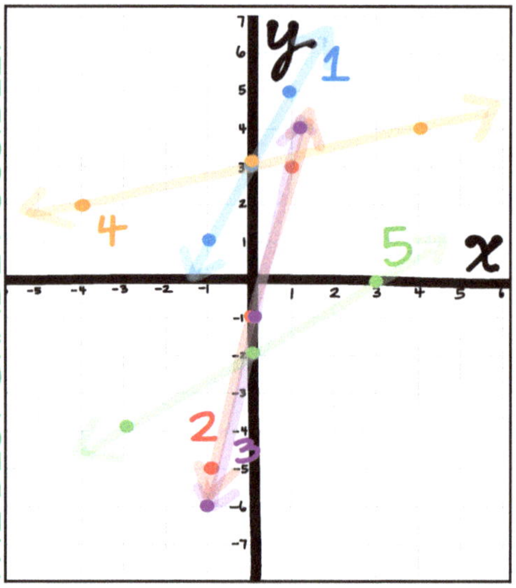

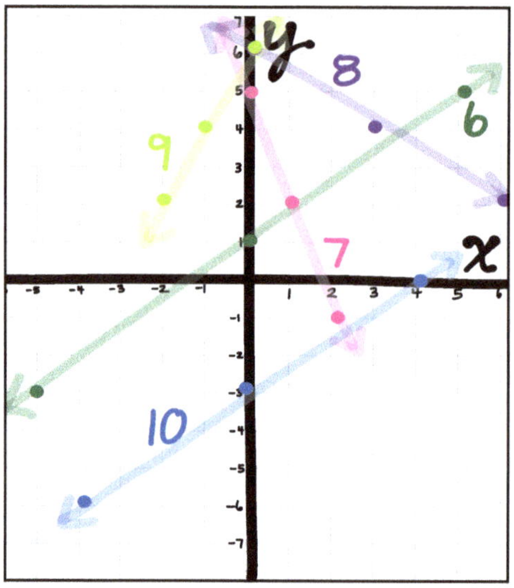

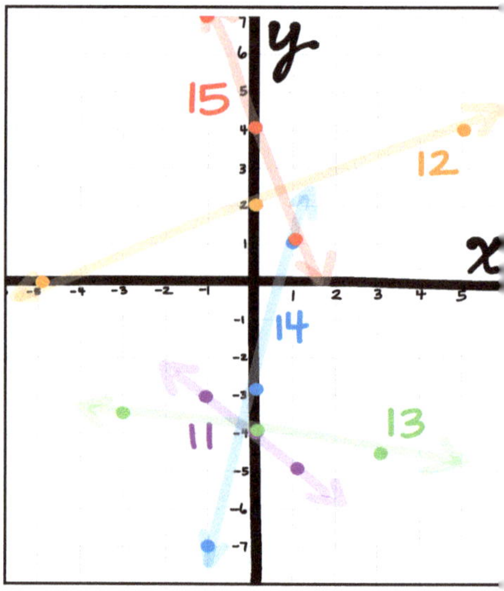

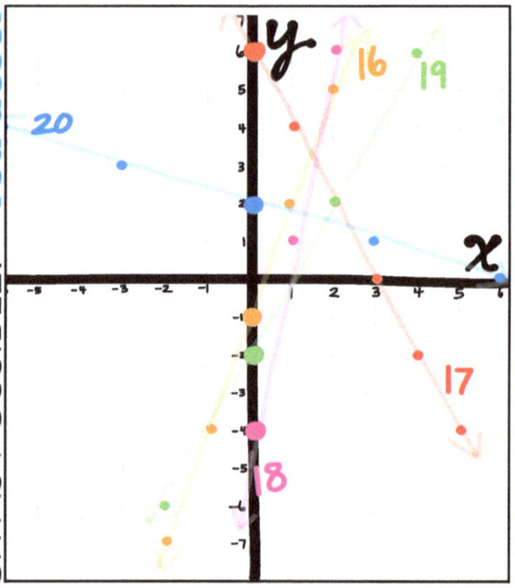

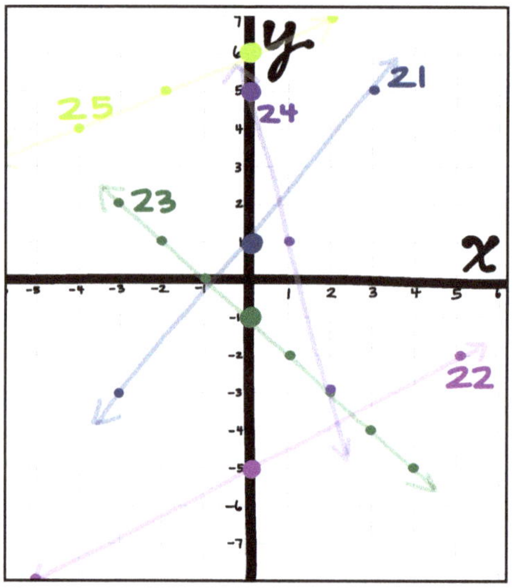

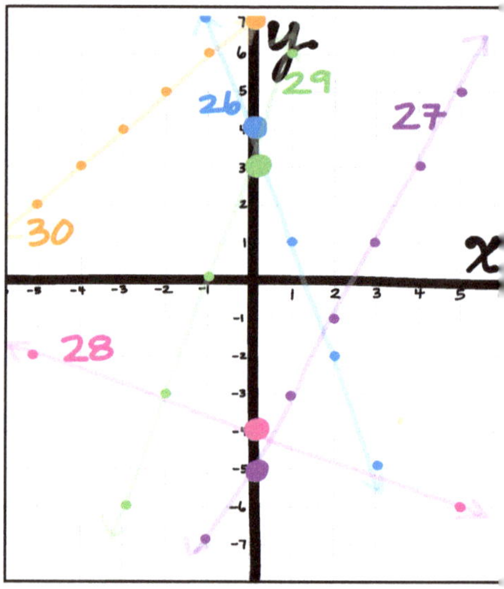

BE AWESOME! Practice AS MUCH AS POSSIBLE! You deserve THE BEST GRADE POSSIBLE!

You = AMAZING

Graphing Horizontal & Vertical Lines
Finding x and y Intercepts

Graph 1 - 15		Find x and y intercepts and graph	
1.	$y = 1$	16.	$8x + 6y = 24$
2.	$y = -4$	17.	$2x - 6y = 12$
3.	$y = 6$	18.	$10x + 8y = 40$
4.	$y = -3$	19.	$-5x + 6y = 30$
5.	$y = -6$	20.	$14x - 7y = 28$
6.	$y = -1$	21.	$-4x - y = 4$
7.	$y = -2$	22.	$7x - 14y = 42$
8.	$x = 6$	23.	$9x - 3y = 18$
9.	$x = -1$	24.	$5x - 5y = 30$
10.	$x = 0$	25.	$2x + 3y = 6$
11.	$x = 3$	26.	$28x - 14y = 56$
12.	$x = -5$	27.	$-3x - 2y = 12$
13.	$x = 4$	28.	$12x + 8y = 48$
14.	$x = -2$	29.	$-8x + 8y = 24$
15.	$x = -3$	30.	$18x + 12y = -36$

Answers: 1-15 and 16-30 see graphs on page 26.

16.	$x = 3$	$y = 4$	26.	$x = 2$	$y = -4$
17.	$x = 6$	$y = -2$	27.	$x = -4$	$y = -6$
18.	$x = 4$	$y = 5$	28.	$x = 4$	$y = 6$
19.	$x = -6$	$y = 5$	29.	$x = -3$	$y = 3$
20.	$x = 2$	$y = -4$	30.	$x = -2$	$y = -3$
21.	$x = -1$	$y = -4$			
22.	$x = 6$	$y = -3$			
23.	$x = 2$	$y = -6$			
24.	$x = 6$	$y = -6$			
25.	$x = 3$	$y = 2$			

Graphing Horizontal and Vertical Lines, Finding x and y Intercepts Answers:

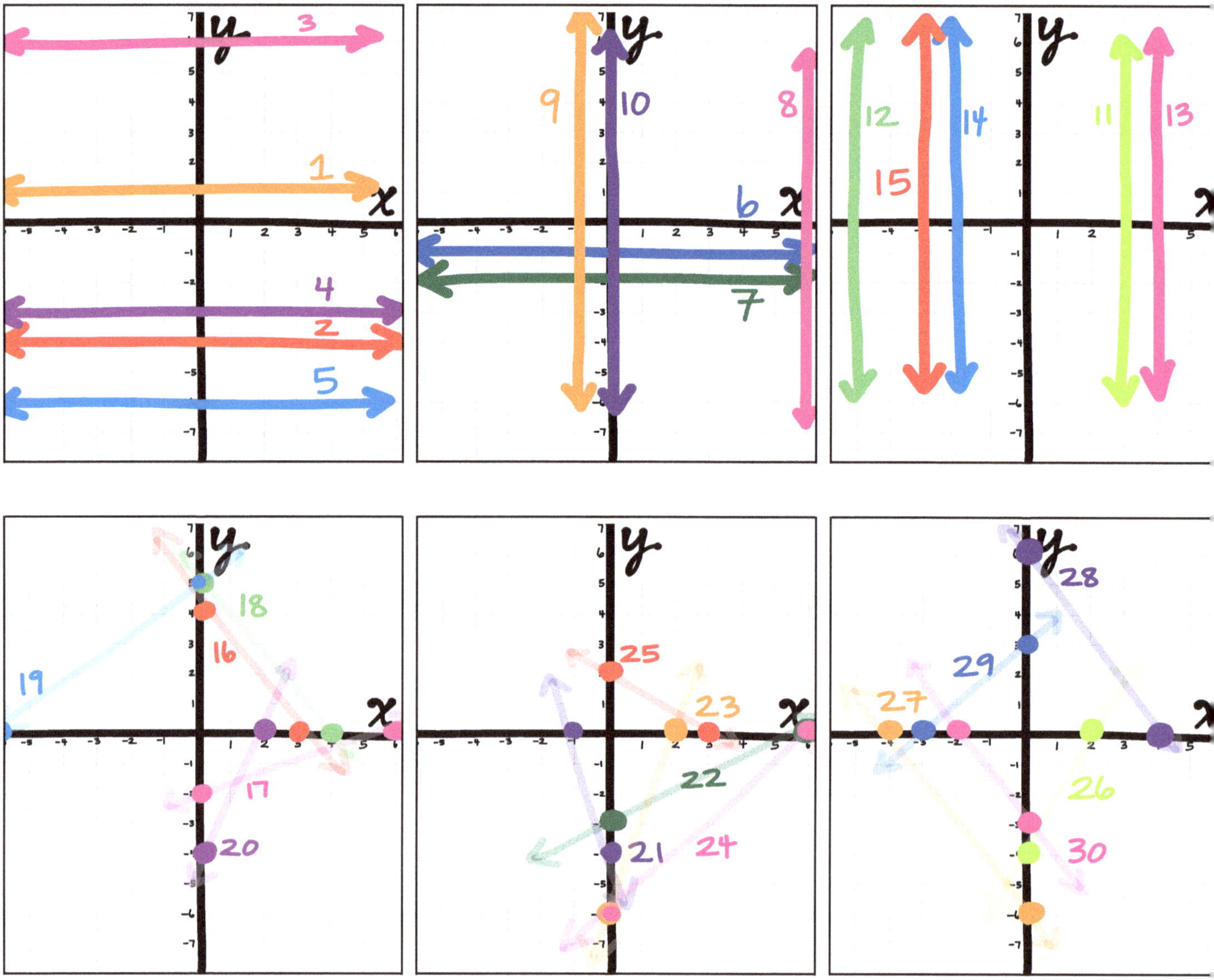

Finding SLOPE from two points
Slope-Intercept Form $y = mx + b$

<u>Find slope given the points</u>

1. (0, 3) (5, 4)
2. (-2, 8) (3, -2)
3. (2, 3) (4, 6)
4. (-1, -4) (3, 8)
5. (-4, 3) (5, -2)
6. (0, 0) (3, 4)
7. (4, 5) (6, 9)
8. (-4, 7) (-6, -4)
9. (7, 3) (-4, 5)
10. (0, 6) (2, 10)
11. (1, 2) (3, 4)
12. (3, 4) (0, -2)
13. (-2, -1) (0, 1)
14. (2, 4) (5, 7)
15. (-5, -6) (2, 8)

<u>Rewrite in slope-intercept form.</u>

16. $8x + 3y = 24$
17. $2x - 6y = 12$
18. $4x + 8y = 40$
19. $5x + 6y = 30$
20. $2x - 7y = 28$
21. $-4x - y = 8$
22. $7x + 3y = 42$
23. $9x - 3y = 18$
24. $5x - 5y = 35$
25. $x + 3y = 9$
26. $8x + 7y = 56$
27. $-3x - 2y = 12$
28. $12x - 8y = 48$
29. $-2x + 4y = 24$
30. $9x + 4y = -36$

Answers

1. $m = 1/5$	11. $m = 1$	21. $y = -4x - 8$
2. $m = -2$	12. $m = 2$	22. $y = -7/3\,x + 14$
3. $m = 3/2$	13. $m = 1$	23. $y = 3x - 6$
4. $m = 3$	14. $m = 1$	24. $y = x - 7$
5. $m = -5/9$	15. $m = 2$	25. $y = -1/3\,x + 3$
6. $m = 4/3$	16. $y = -8/3\,x + 8$	26. $y = -8/7\,x + 8$
7. $m = 2$	17. $y = 1/3\,x - 2$	27. $y = -3/2\,x - 6$
8. $m = 11/2$	18. $y = -1/2\,x + 5$	28. $y = 3/2\,x - 6$
9. $m = -2/11$	19. $y = -5/6\,x + 5$	29. $y = 1/2\,x + 6$
10. $m = 2$	20. $y = 2/7\,x - 4$	30. $y = -9/4\,x - 9$

Algebra is EASY and YOU CAN DO IT

Point-Slope Form $y - y_1 = m(x - x_1)$

BE AWESOME! Practice AS MUCH AS POSSIBLE! You deserve THE BEST GRADE POSSIBLE!

Write the equation of the line containing the given point and slope in point-slope form.

1.	(2, 3)	m = 3
2.	(0, -1)	m = -2
3.	(-6, 9)	m = 1
4.	(3, 7)	m = $\frac{1}{2}$
5.	(4, 0)	m = -1
6.	(-1, 5)	m = $\frac{2}{3}$
7.	(7, -4)	m = -4
8.	(0, 6)	m = $\frac{3}{4}$
9.	(1, -3)	m = 2
10.	(-4, 5)	m = 1
11.	(-2, -2)	m = 3
12.	(-3, 5)	m = $\frac{5}{2}$
13.	(0, -2)	m = -1
14.	(3, 0)	m = $-\frac{1}{3}$
15.	(1, -3)	m = 4

Write the equation of the line containing the the given points in point-slope form.

16. (0, 2) and (1, -3)
17. (4, -2) and (-4, -4)
18. (-4, 5) and (5, -5)
19. (0, 3) and (-4, -1)
20. (5, -2) and (-4, -3)
21. (-5, -3) and (0, -2)
22. (1, 5) and (-8, 0)
23. (5, 4) and (-4, 3)
24. (0, -1) and (-3, 4)
25. (5, 0) and (6, 2)
26. (-1, -1) and (6, 6)
27. (8, 2) and (8, -5)
28. (7, -1) and (9, -3)
29. (0, 4) and (3, 4)
30. (-6, 7) and (0, 0)

Answers: 16-30 use the first given point plugged into equation

1. $y - 3 = 3(x - 2)$
2. $y + 1 = -2(x - 0)$
3. $y - 9 = 1(x + 6)$
4. $y - 7 = 1/2\,(x - 3)$
5. $y - 0 = -1(x - 4)$
6. $y - 5 = 2/3\,(x + 1)$
7. $y + 4 = -4(x - 7)$
8. $y - 6 = 3/4\,(x - 0)$
9. $y + 3 = 2(x - 1)$
10. $y - 5 = 1(x + 4)$
11. $y + 2 = 3(x + 2)$
12. $y - 5 = 5/2\,(x + 3)$
13. $y + 2 = -1(x - 0)$
14. $y - 0 = -1/3\,(x - 3)$
15. $y + 3 = 4(x - 1)$
16. $y - 2 = -5(x - 0)$
17. $y + 2 = 1/4\,(x - 4)$
18. $y - 5 = -10/9\,(x + 4)$
19. $y - 3 = 1(x - 0)$
20. $y + 2 = 1/9\,(x - 5)$
21. $y + 3 = 1/5\,(x + 5)$
22. $y - 5 = 5/9\,(x - 1)$
23. $y - 4 = 1/9\,(x - 5)$
24. $y + 1 = -5/3\,(x - 0)$
25. $y - 0 = -2(x - 5)$
26. $y + 1 = 1(x + 1)$
27. $x = 8$, NO SLOPE
28. $y + 1 = -1(x - 7)$
29. $y - 4 = 0(x - 0)$, $y = 4$
30. $y - 7 = -7/6\,(x + 6)$

You = AMAZING

Parallel and Perpendicular Lines

Write the equation of the line that passes through the given point and is parallel to the given line (answer in point slope form)

1.	(0, 6)	$y = 4x - 1$
2.	(-1, 2)	$y = -2x + 1$
3.	(5, -4)	$y = 8 - 4x + 6x$
4.	(-6, 0)	$y = \frac{1}{2}x - 3$
5.	(3, 4)	$y = -x + 5$
6.	(4, 8)	$2x + 4y = 12$
7.	(-1, 5)	$9x - 3y = 21$
8.	(-3, 7)	$5x + 4y = 12$
9.	(2, 3)	$x - 2y = 6$
10.	(0, -4)	$-3x + 9y = -6$
11.	(3, -2)	$y = 5x - 1$
12.	(4, 5)	$8x - 2y = 14$
13.	(-1, -1)	$3x + 7y = 11$
14.	(0, -2)	$-x + 3y = 6$
15.	(-5, 3)	$6x + 4y = 10$

Write the equation of the line that passes through the given point and is perpendicular to the given line (answer in point slope form)

16.	(3, -5)	$4x + 6y = 20$
17.	(-2, 0)	$3x - y = 6$
18.	(1, -1)	$6x + 3y = 12$
19.	(5, 4)	$2x - 8y = 16$
20.	(2, -3)	$y = x - 5$
21.	(4, 0)	$9x - 3y = 12$
22.	(-3, 2)	$2x - y = -5$
23.	(-7, 3)	$4x - 5y = 20$
24.	(-5, 1)	$3x - 9y = 18$
25.	(8, -4)	$4x - 2y = 8$
26.	(-4, 3)	$y = \frac{1}{2}x + 3$
27.	(0, 6)	$y = \frac{3}{4}x - 1$
28.	(-2, 1)	$4x - 4y = 7$
29.	(7, 0)	$-2x + 5y = 9$
30.	(-1, -2)	$y = \frac{-2}{3}x + 5$

Answers:

1. $y - 6 = 4(x - 0)$
2. $y - 2 = -2(x + 1)$
3. $y + 4 + 2(x - 5)$
4. $y - 0 = 1/2\ (x + 6)$
5. $y - 4 = -1(x - 3)$
6. $y - 8 = -1/2\ (x - 4)$
7. $y - 5 = 3(x + 1)$
8. $y - 7 = -5/4\ (x + 3)$
9. $y - 3 = 1/2\ (x - 2)$
10. $y + 4 = 1/3\ (x - 0)$
11. $y + 2 = 5(x - 3)$
12. $y - 5 = 4\ (x - 4)$
13. $y + 1 = -3/7\ (x + 1)$
14. $y + 2 = 1/3\ (x - 0)$
15. $y - 3 = -3/2\ (x + 5)$
16. $y + 5 = 3/2\ (x - 3)$
17. $y - 0 = -1/3\ (x + 2)$
18. $y + 1 = 1/2\ (x - 1)$
19. $y - 4 = 4(x - 5)$
20. $y + 3 = -1(x - 2)$
21. $y - 0 = 1/3\ (x - 4)$
22. $y - 2 = -1/2\ (x + 3)$
23. $y - 3 = -5/4\ (x + 7)$
24. $y - 1 = -3(x + 5)$
25. $y + 4 = -1/2\ (x - 8)$
26. $y - 3 = -2(x + 4)$
27. $y - 6 = -4/3\ (x - 0)$
28. $y - 1 = -1(x + 2)$
29. $y - 0 = -5/2\ (x - 7)$
30. $y + 2 = 3/2\ (x + 1)$

Step by step: See each problem in SUCCESS BOOK solved step-by-step in WEBINAR

Algebra is EASY and YOU CAN DO IT

Inequalities, ± Equations
Inequalities, x/÷ Equations

Practice AS MUCH AS POSSIBLE! You deserve THE BEST GRADE POSSIBLE!

Solve the following inequalities:

1. $x - 5 > 12$
2. $x + 8 < 16$
3. $x - 3 \geq 0$
4. $8 + x \leq -6$
5. $x + 7 > -9$
6. $x > 7 - 12$
7. $9 + x < -9$
8. $12 \geq x - 4$
9. $7x - 3 < 8x - 1$
10. $11x + 5 > 12x$
11. $x - 8 < 9 - 21$
12. $5 + x > -3$
13. $x + 16 \geq -1$
14. $x - 4 < -4$
15. $3 + x > 7 + 12$
16. $\frac{5}{2} x > 3$
17. $3x < -3$
18. $-5x \leq 15$
19. $8x > -24$
20. $-\frac{2}{3} x < -1$
21. $4 \leq \frac{4}{5} x$
22. $7x > 21$
23. $-2x < 18$
24. $-x < 9$
25. $4x > 16$
26. $\frac{3}{7} x > 8$
27. $6x < 2$
28. $-5x \geq 25$
29. $8x < -64$
30. $\frac{3}{2} x < 9$

BE AWESOME!

Answers:

1. $x > 17$
2. $x < 8$
3. $x \geq 3$
4. $x \leq 14$
5. $x > -16$
6. $x > -5$
7. $x < -18$
8. $x \leq 16$
9. $x < 2$
10. $x < 5$
11. $x < -4$
12. $x > -8$
13. $x \geq -17$
14. $x < 0$
15. $x > 16$
16. $x > 6/5$
17. $x < -1$
18. $x \geq -3$
19. $x \geq -3$
20. $x > 3/2$
21. $x \geq 5$
22. $x > 3$
23. $x > -9$
24. $x > -9$
25. $x > 4$
26. $x > 56/3$
27. $x < 1/3$
28. $x \leq -5$
29. $x < -8$
30. $x < 6$

You = AMAZING

Inequalities, Multi-Step Equations
Compound Inequalities

Solve the following multi-step inequalities:

1. $x + 4 > 9x - 5$
2. $3x + 8x > -14$
3. $7 - x + 9 \le 6$
4. $-4x + 8 < 7x + 9$
5. $3(x + 2) \ge 18$
6. $-5(2x - 4) > 2(x + 9)$
7. $12 - 3(-2x - 1) \le 20$
8. $5x - 4 < 8x + 3$
9. $2(6x + 1) \ge 3x - 5$
10. $7x > 8x + 3(4x -3)$
11. $-6x - 7 < 18 - 12 + 4x$
12. $14x - 5 \ge -25$
13. $3(2x - 5) > -20x$
14. $25x < 18x - 14$
15. $4(5x + 2) \ge 9(2x - 1)$

Solve the following compound inequalities:

16. $-3 < x - 2 < 8$
17. $4 < x + 3 < 16$
18. $-5 \le 2x - 5 < 11$
19. $0 < 5x < 25$
20. $-2 < 6x - 3 \le 33$
21. $-12 < 3x + 9 < 21$
22. $25 \le -5x \le 50$
23. $12 < 7x + 5 < 19$
24. $4 < 9 - 5x \le 24$
25. $8 < 24x < 16$
26. $-5 \le 11x + 6 \le 28$
27. $2 < 4(2x - 3) < 8$
28. $13 \le 18x - 5 \le 31$
29. $6 < 2x + 8 \le 12$
30. $-4 < -7x + 3 < 10$

Answers

1.	$x < 9/8$	11.	$x > -13/10$	21.	$-7 < x < 4$
2.	$x > -14/5$	12.	$x \ge 10/7$	22.	$-10 \le x \le -5$
3.	$x \ge 10$	13.	$x > 15/26$	23.	$1 < x < 2$
4.	$x > -1/11$	14.	$x < -2$	24.	$-3 \le x < 1$
5.	$x \ge 4$	15.	$x \ge -17/2$	25.	$1/3 < x < 2/3$
6.	$x < 1/6$	16.	$-1 < x < 10$	26.	$-1 \le x \le 2$
7.	$x \le 5/6$	17.	$1 < x < 13$	27.	$7/4 < x < 5/2$
8.	$x > -7/3$	18.	$0 \le x < 8$	28.	$1 \le x \le 2$
9.	$x \ge -7/9$	19.	$0 < x < 5$	29.	$-1 < x \le 2$
10.	$x < 9/13$	20.	$1/6 < x \le 6$	30.	$-1 < x < 1$

step by step: See each problem in SUCCESS BOOK solved step-by-step in WEBINAR

Absolute Value Inequalities

Solve the following inequalities:

1. $|x| > 3$
2. $|x| < 12$
3. $|x + 2| > 3$
4. $|x - 4| \leq 5$
5. $|4x| < 16$
6. $|8x| > 12$
7. $3|x - 1| \geq 12$
8. $2|x + 3| < 22$
9. $|9x - 5| \geq 13$
10. $|4x + 7| < 21$
11. $6|2x - 9| > 42$
12. $3|3x + 2| < 21$
13. $2|x + 5| - 4 \leq 20$
14. $|5x - 3| + 6 > 12$
15. $3|4x + 4| -7 \leq 14$

Graph the following on a number line:

16. $x > 5$
17. $x < -2$
18. $x \geq 0$
19. $x \leq -6$
20. $x < 5$
21. $x > -5$
22. $-2 < x < 1$
23. $2 < x < 6$
24. $-6 \leq x < 2$
25. $0 < x < 4$
26. $-1 < x < 5$
27. $-3 < x < 6$
28. $x < 2$ and $x > 6$
29. $x < -3$ and $x \geq 0$
30. $x < -5$ and $x > 4$

Answers: 16-30 graphs on page 34

1.	$x > 3$	$x < -3$
2.	$x < 12$	$x > -12$ or $-12 < x < 12$
3.	$x > 1$	$x < -5$
4.	$x \leq 9$	$x \geq -1$ or $-1 \leq x \leq 9$
5.	$x < 4$	$x > -4$ or $-4 < x < 4$
6.	$x > 3/2$	$x < -3/2$
7.	$x \geq 5$	$x \leq -3$
8.	$x < 8$	$x > -14$ or $-14 < x < 8$
9.	$x \geq 2$	$x \leq -8/9$
10.	$x < 7/2$	$x > -7$ or $-7 < x < 7/2$
11.	$x > 8$	$x < -1$
12.	$x < 5/3$	$x > -3$ or $-3 < x < 5/3$
13.	$x \leq 7$	$x \geq -17$ or $-17 \leq x \leq 7$
14.	$x > 9/5$	$x < -3/5$
15.	$x \leq 3/4$	$x \geq -11/4$ or $-11/4 \leq x \leq 3/4$

BE AWESOME! Practice AS MUCH AS POSSIBLE! You deserve THE BEST GRADE POSSIBLE!

You = AMAZING

Graphing Inequalities on a Grid

Graph the following inequalities

1. $y \geq 6$
2. $y < -2$
3. $x > 1$
4. $x \leq -3$
5. $y > 3$
6. $x \leq 0$ and $y > 3$
7. $x \geq -1$ and $y < 1$
8. $x < 5$ and $y \geq 0$
9. $y \leq -5$ and $x \geq 3$
10. $y > 2$ and $x < -4$
11. $y < x + 1$
12. $y \leq -x - 4$
13. $y \geq 2x - 5$
14. $y < \frac{-2}{3}x + 2$
15. $y > \frac{1}{4}x - 3$

Graph the following on a grid.

16. $y < 3x - 1$
    $y > -\frac{2}{3}x + 4$
17. $y \geq x$
    $y < 4x + 5$
18. $y < -2x + 3$
    $y \leq \frac{5}{4}x - 2$
19. $7x - 3y < 21$
    $4x + 3y \geq 12$
20. $-3x - 2y < -8$
    $5x + y < 1$
21. $x + y > -2$
    $-2x + y < -5$
22. $y \geq -3x + 4$
    $y < \frac{3}{5}x - 5$

step by step: See each problem in SUCCESS BOOK solved step-by-step in WEBINAR

Answers on page 34, 35

Absolute Value Inequalities Answers 16 - 30

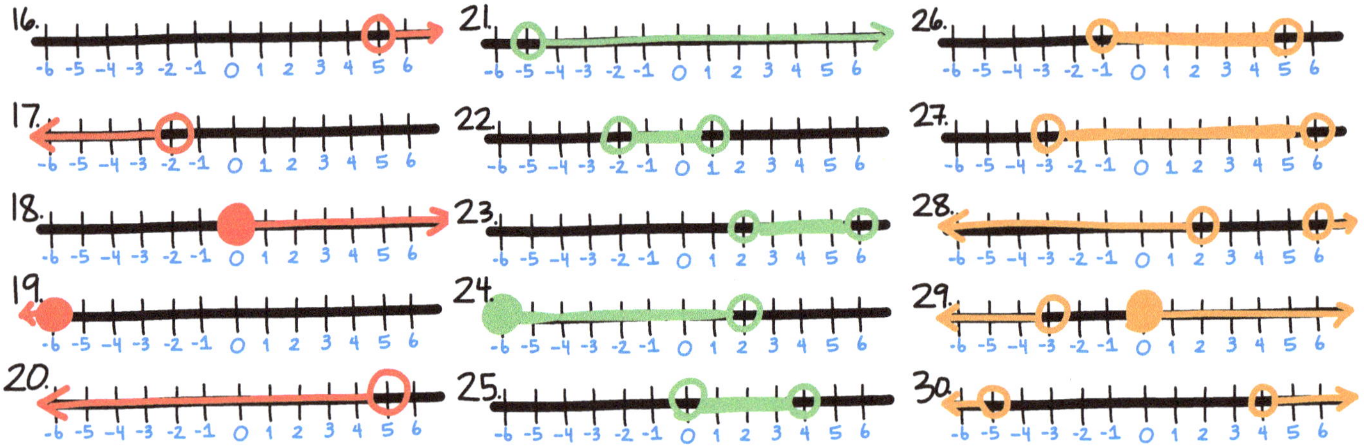

Graphing Inequalities on a Grid Answers

BE AWESOME! Practice AS MUCH AS POSSIBLE! You deserve THE BEST GRADE POSSIBLE!

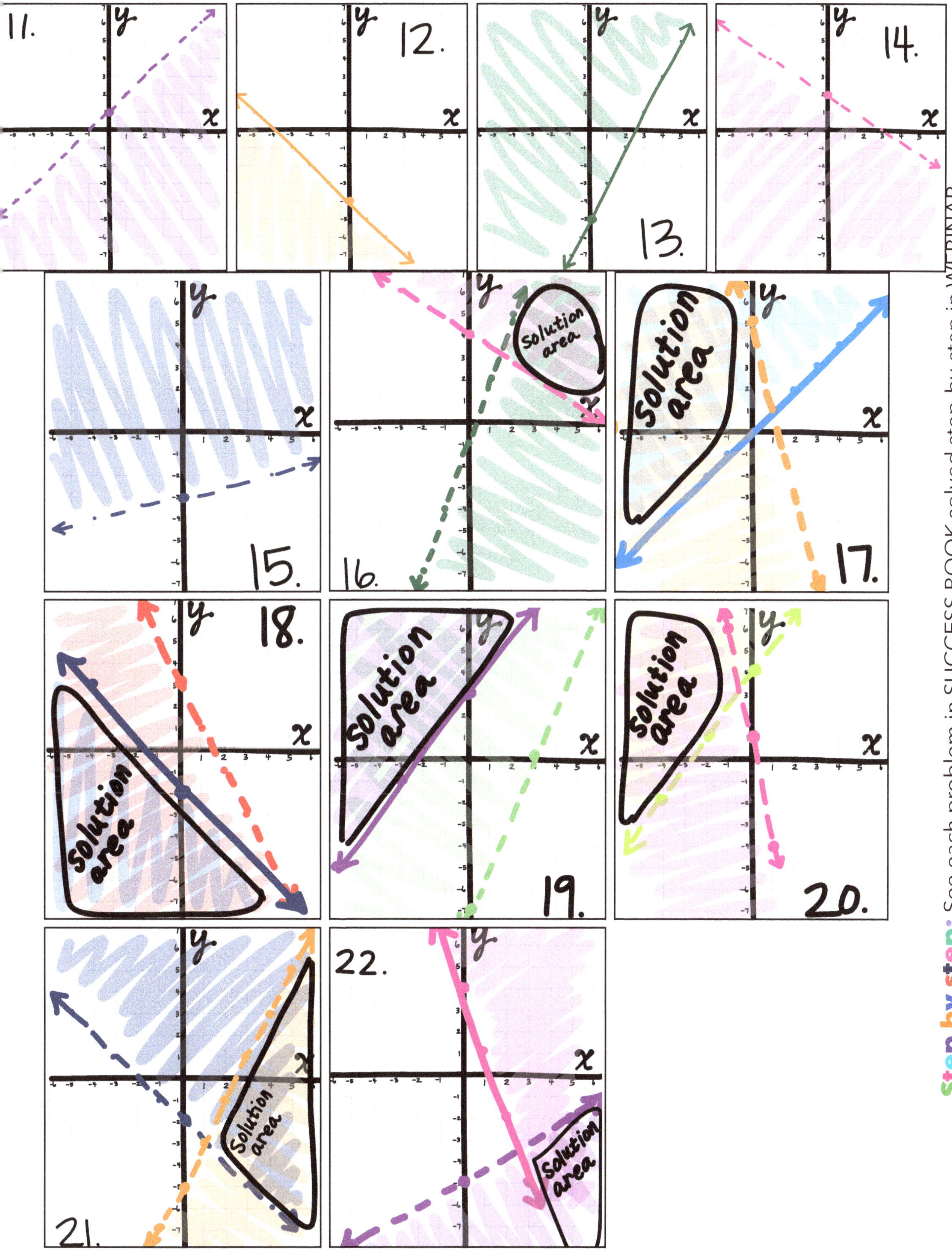

Step by step: See each problem in SUCCESS BOOK solved step-by-step in WEBINAR

BASICS Test #1

Solve the following:

1. $8 + (-9) =$ 2. $12 - 17 =$ 3. $-5 - 4 =$

4. $-18 + (-2) =$ 5. $9 + (-12) - 18 =$ 6. $3 - 5 - 8 =$

7. $26 - 32 =$ 8. $20 + (-5) =$ 9. $-5 - (-12) =$

10. $25 \bullet 2 =$ 11. $26 \bullet -32 =$ 12. $-10 \bullet 10 =$

13. $65 \div -5 =$ 14. $(-3)(-2)(-4) =$ 15. $9 \times 15 =$

16. $125 \div -25 =$ 17. $-8 \bullet (-3) =$ 18. $90 \div 15 =$

19. $5 - 4(35 \div 5)^2 =$ 20. $(-4)^3 + 5 - 3 \times 8 =$

21. $12 - 4 \times 9 =$ 22. $100 + 3 \times 9 - 25 \div -5 =$

23. $8(121 \div -11 + 14) - 5^3 =$ 24. $-20 \bullet -4 + 3^3 - 10 =$

25. $8(9 + 12) =$ 26. $-4(12 \div -2) =$ 27. $5(x + 2) =$

28. $-3(y - 2) =$ 29. $x(x + 4) =$ 30. $y^3(x + y) =$

31. $x^4(3 - x) =$ 32. $a^2(9 + a) =$ 33. $2c(c^2 - c) =$

Evaluate the following for x = -2, y = 3 and z = -5

34. $3x^2 =$ 35. $9y + x^3 =$ 36. $x + 3y - z =$

37. $x(x - z) =$ 38. $z^2 + z =$ 39. $z(x^2 - y^2) =$

40. $18x + 9y - z =$ 41. $z^3(z^4 + z) =$ 42. $x - y + 5z =$

Solve the following:

43. $|-5| =$ 44. $|12 - 25| =$ 45. $-|-25| =$

Simplify the following expressions:

46. $22x + 5x =$ 47. $9 + 12y - 6 + 3y =$ 48. $11x + 7y - 6x =$

49. $3x^2 + 7y^3 - 12x^2 + 25y^3 =$ 50. $5abc + 7abc - 18abc =$

51. $12x(x + 3x^2) =$ 52. $8a \bullet 3a^3 \bullet 2a^2 =$ 53. $4y^2(y^2 + 5y^3 - 7y^5) =$

54. $2a^3b^4(4ab - 6a^4b^5) =$ 55. $65x^3 \div 5x =$

56. $\frac{25x^2y^5}{5xy^3} =$ 57. $\frac{18a^5bc^3}{9ab^4} =$ 58. $\frac{7x^7y^3z^2}{21x^3yz^5} =$

BE AWESOME! Practice AS MUCH AS POSSIBLE! You deserve THE BEST GRADE POSSIBLE!

BASICS Test #2

Solve the following:

1. $12 + (-7) =$
2. $14 - 19 =$
3. $-8 - 5 =$
4. $-20 + (-5) =$
5. $6 + (-11) - 10 =$
6. $4 - 7 - 4 =$
7. $25 - 39 =$
8. $22 + (-8) =$
9. $-9 - (-11) =$
10. $26 \bullet -3 =$
11. $22 \bullet -12 =$
12. $-15 \bullet 10 =$
13. $95 \div -5 =$
14. $(-4)(-1)(-5) =$
15. $8 \times 16 =$
16. $186 \div -2 =$
17. $-5 \bullet (-3) =$
18. $-60 \div 15 =$
19. $7 - 2(45 \div 5)^2 =$
20. $(-3)^3 + 8 - 4 \times 7 =$
21. $22 - 8 \bullet (-9) =$
22. $100 + 3 \bullet 9 - 40 \div -5 =$
23. $9(81 \div -3 + 11) - 4^3 =$
24. $-25 \bullet -2 + 5^3 - 11 =$
25. $12(5 + 17) =$
26. $-3(20 \div -2) =$
27. $9(x + 7) =$
28. $-2(5y - 6) =$
29. $x^2(2x + 6) =$
30. $x^3(x + 5y) =$
31. $x^4(3x^2 - 4x) =$
32. $a^2(9b + 5a^2) =$
33. $8c(c^2 - c) =$

Evaluate the following for x = -4, y = -1 and z = 2

34. $6x^3 =$
35. $5y^2 + x^3 =$
36. $2x + 4y - 2z =$
37. $y(x - z) =$
38. $z^2 + x^4 =$
39. $y(y^2 - z^2) =$
40. $11x - 8y + z =$
41. $y^4(z^3 + x) =$
42. $6x - 5y + 6z =$

Solve the following:

43. $|-8| =$
44. $|11 - 20| =$
45. $-|-75| =$

Simplify the following expressions:

46. $7x + 9x =$
47. $5 + 12y - 4 + 12y =$
48. $15x + 21y - 9x =$
49. $5a^2 + 8b^3 - 42a^2 + 15b^3 =$
50. $7abc + 21abc - 17abc =$
51. $15x(2x^3 + 3x^2) =$
52. $9a^2 \bullet 4a^3 \bullet 5a^2 =$
53. $3y^3(y^4 + 5y^3 - 7y^5) =$
54. $5a^3b^4(3ab - 7a^4b^5) =$
55. $15x^3 \div 3x =$
56. $\frac{18x^2y^6}{6x^5y^3} =$
57. $\frac{75a^5bc^3}{5a^3b^4} =$
58. $\frac{9x^6y^2z}{81x^4yz^6} =$

Step by step: See each problem in SUCCESS BOOK solved step-by-step in WEBINAR

Algebra is EASY and YOU CAN DO IT

Basics Test #1 Answers:

1. -1
2. -5
3. -9
4. -20
5. -21
6. -10
7. -6
8. 15
9. 7
10. 50
11. -832
12. -100
13. -13
14. -24
15. 135
16. -5
17. 24
18. 6
19. -191
20. -83
21. -24
22. 132
23. -101
24. 97
25. 168
26. 24
27. $5x + 10$
28. $-3y + 6$
29. $x^2 + 4x$
30. $xy^3 + y^4$
31. $3x^4 - x^5$
32. $9a^2 + a^3$
33. $2c^3 - 2c^2$
34. 12
35. 19
36. 12
37. -6
38. 20
39. 25
40. -4
41. $-77,500$
42. -30
43. 5
44. 13
45. -25
46. $27x$
47. $15y + 3$
48. $5x + 7y$
49. $-9x^2 + 32y^3$
50. $-6abc$
51. $12x^2 + 36x^3$
52. $48a^6$
53. $4y^4 + 20y^5 - 28y^7$
54. $8a^4b^5 - 12a^7b^9$
55. $13x^2$
56. $5xy^2$
57. $\frac{2a^4c^3}{b^3}$
58. $\frac{x^4y^2}{3z^3}$

BE AWESOME! Practice AS MUCH AS POSSIBLE! You deserve THE BEST GRADE POSSIBLE!

You = AMAZING

Basics Test #2 Answers:

1. 5
2. -5
3. -13
4. -25
5. -15
6. -7
7. -14
8. 14
9. 2
10. -78
11. -264
12. -140
13. -19
14. -20
15. 128
16. -93
17. 15
18. -4
19. -155
20. -47
21. 94
22. 135
23. -208
24. 164
25. 264
26. 30
27. $9x + 63$
28. $-10y + 12$
29. $2x^3 + 6x^2$
30. $x^4 + 5x^3y$
31. $3x^6 - 4x^5$
32. $9a^2b + 5a^4$
33. $8c^3 - 8c^2$
34. -384
35. -59
36. -16
37. 6
38. 260
39. 3
40. -34
41. 4
42. -7
43. 8
44. 9
45. -75
46. $16x$
47. $24y + 1$
48. $6x + 21y$
49. $-37a^2 + 23b^3$
50. $11abc$
51. $30x^4 + 45x^3$
52. $180a^7$
53. $3y^7 + 15y^6 - 21y^8$
54. $15a^4b^5 - 35a^7b^9$
55. $5x^3$
57. $\frac{3y^3}{x^3}$
57. $\frac{15a^2c^3}{b^3}$
58. $\frac{x^2y}{9z^5}$

Step by step: See each problem in SUCCESS BOOK solved step-by-step in WEBINAR

FRACTIONS Test #1

Simplify the following fractions:

1. $\frac{6}{8} =$ 2. $\frac{9}{81} =$ 3. $\frac{14}{21} =$

4. $\frac{12}{18} =$ 5. $\frac{6}{38} =$ 6. $\frac{2}{90} =$

7. $\frac{20}{65} =$ 8. $\frac{32}{64} =$ 9. $\frac{19}{38} =$

10. $\frac{18}{28} =$ 11. $\frac{7}{42} =$ 12. $\frac{15}{40} =$

Are the following fractions in lowest terms? If so, write yes. If not, simplify.

13. $\frac{2}{9} =$ 14. $\frac{8}{24} =$ 15. $\frac{9}{80} =$

Solve, then simplify to lowest terms (answer as Mixed Number, no improper fractions):

16. $\frac{1}{4} + \frac{7}{8} =$ 17. $\frac{9}{10} - \frac{2}{5} =$ 18. $3\frac{1}{4} + 4\frac{2}{3} =$

19. $\frac{6}{7} - \frac{2}{9} =$ 20. $5\frac{1}{5} - 2\frac{3}{4} =$ 21. $12\frac{8}{9} - 1\frac{9}{10} =$

Convert the following improper fractions into mixed numbers, simplify.

22. $\frac{9}{5} =$ 23. $\frac{12}{3} =$ 24. $\frac{18}{4} =$

25. $\frac{28}{6} =$ 26. $\frac{90}{8} =$ 27. $\frac{25}{6} =$

Convert the following mixed numbers into improper fractions, simplify.

28. $4\frac{1}{3} =$ 29. $8\frac{3}{8} =$ 30. $12\frac{3}{4} =$

31. $1\frac{2}{5} =$ 32. $3\frac{7}{9} =$ 33. $6\frac{1}{4} =$

Solve:

34. $\frac{1}{2} \bullet \frac{4}{5} =$ 35. $\frac{8}{9} \bullet \frac{2}{3} =$ 36. $\frac{3}{4} \bullet \frac{1}{2} =$

37. $3\frac{1}{3} \bullet 1\frac{1}{2} =$ 38. $5\frac{2}{7} \bullet 2\frac{3}{5} =$ 39. $1\frac{2}{9} \bullet 1\frac{2}{3} =$

40. $\frac{1}{7} \div \frac{2}{3} =$ 41. $\frac{3}{4} \div \frac{9}{10} =$ 42. $\frac{5}{6} \div \frac{1}{2} =$

43. $1\frac{4}{7} \div 2\frac{3}{4} =$ 44. $5\frac{1}{3} \div 3\frac{4}{5} =$ 45. $8\frac{2}{3} \div 2\frac{9}{10} =$

Order from least to greatest:

46. $\frac{7}{8}, 1, \frac{3}{4} =$ 47. $\frac{1}{2}, \frac{3}{5}, \frac{2}{3} =$ 48. $\frac{5}{6}, \frac{3}{4}, \frac{7}{8} =$

BE AWESOME! Practice AS MUCH AS POSSIBLE! You deserve THE BEST GRADE POSSIBLE!

You = AMAZING

FRACTIONS Test #2

Simplify the following fractions:

1. $\frac{5}{25} =$	2. $\frac{24}{81} =$	3. $\frac{11}{55} =$
4. $\frac{3}{39} =$	5. $\frac{16}{48} =$	6. $\frac{3}{81} =$
7. $\frac{12}{60} =$	8. $\frac{3}{63} =$	9. $\frac{8}{44} =$
10. $\frac{14}{38} =$	11. $\frac{13}{39} =$	12. $\frac{5}{95} =$

Are the following fractions in lowest terms? If so, write yes. If not, simplify.

13. $\frac{12}{15} =$	14. $\frac{3}{24} =$	15. $\frac{7}{9} =$

Solve, then simplify to lowest terms (answer as Mixed Number, no improper fractions):

16. $\frac{2}{5} + \frac{7}{9} =$	17. $\frac{5}{12} - \frac{7}{10} =$	18. $2\frac{3}{7} + 3\frac{1}{3} =$
19. $\frac{5}{6} - \frac{3}{5} =$	20. $3\frac{3}{4} - 1\frac{1}{5} =$	21. $5\frac{1}{3} - 2\frac{3}{4} =$

Convert the following improper fractions into mixed numbers, simplify.

22. $\frac{11}{5} =$	23. $\frac{14}{3} =$	24. $\frac{25}{4} =$
25. $\frac{22}{6} =$	26. $\frac{45}{8} =$	27. $\frac{20}{3} =$

Convert the following mixed numbers into improper fractions, simplify.

28. $3\frac{2}{5} =$	29. $5\frac{2}{7} =$	30. $10\frac{2}{3} =$
31. $2\frac{6}{7} =$	32. $4\frac{3}{11} =$	33. $5\frac{5}{6} =$

Solve:

34. $\frac{2}{3} \bullet \frac{5}{2} =$	35. $\frac{8}{9} \bullet \frac{9}{8} =$	36. $\frac{1}{4} \bullet \frac{2}{5} =$
37. $5\frac{2}{3} \bullet 2\frac{1}{2} =$	38. $2\frac{3}{7} \bullet 1\frac{4}{5} =$	39. $4\frac{1}{4} \bullet 3\frac{1}{3} =$
40. $\frac{4}{7} \div \frac{2}{5} =$	41. $\frac{1}{4} \div \frac{7}{10} =$	42. $\frac{8}{9} \div \frac{2}{3} =$
43. $3\frac{1}{6} \div 1\frac{1}{4} =$	44. $1\frac{5}{11} \div 2\frac{6}{7} =$	45. $4\frac{2}{3} \div 1\frac{7}{8} =$

Order from least to greatest:

46. $1\frac{5}{8},\ 1\frac{1}{2},\ 1\frac{3}{5} =$	47. $\frac{4}{5},\ \frac{5}{6},\ \frac{6}{7} =$	48. $\frac{1}{2},\ \frac{1}{3},\ \frac{1}{4} =$

Step by step: See each problem in SUCCESS BOOK solved step-by-step in WEBINAR

Fractions Test #1 Answers:

1. 3/4
2. 1/9
3. 2/3
4. 2/3
5. 3/19
6. 1/45
7. 4/13
8. 1/2
9. 1/2
10. 9/14
11. 1/6
12. 3/8
13. YES
14. NO 1/3
15. YES
16. 9/8 or 1 1/8
17. 1/2
18. 7 11/12
19. 40/63
20. 2 9/20
21. 10 89/90
22. 1 4/5
23. 4
24. 4 1/2
25. 4 2/3
26. 11 1/4
27. 4 1/6
28. 13/3
29. 67/8
30. 51/4
31. 7/5
32. 34/9
33. 25/4
34. 2/5
35. 16/27
36. 3/8
37. 5
38. 13 26/35
39. 2 1/27
40. 3/14
41. 5/6
42. 1 2/3
43. 4/7
44. 1 23/57
45. 80/87
46. 3/4, 7/8, 1
47. 1/2, 3/5, 2/3
48. 3/4, 5/6, 7/8

Fractions Test #2 Answers:

1. 1/5
2. 8/27
3. 1/5
4. 1/13
5. 1/3
6. 1/27
7. 1/5
8. 1/21
9. 2/11
10. 7/19
11. 1/3
12. 1/19
13. NO 4/5
14. NO 1/8
15. YES
16. 1 8/45
17. -17/60
18. 5 16/21
19. 7/30
20. 2 11/20
21. 2 7/12
22. 2 1/5
23. 4 2/3
24. 6 1/4
25. 3 2/3
26. 5 5/8
27. 6 2/3
28. 17/5
29. 37/7
30. 32/3
31. 20/7
32. 47/11
33. 35/6
34. 1 2/3
35. 1
36. 1/10
37. 14 1/6
38. 4 13/35
39. 14 1/6
40. 1 3/7
41. 5/14
42. 1 1/3
43. 2 8/15
44. 28/55
45. 2 22/45
46. 1 1/2, 1 3/5, 1 5/8
47. 4/5, 5/6, 6/7
48. 1/4, 1/3, 1/2

Step by step: See each problem in SUCCESS BOOK solved step-by-step in WEBINAR

EXPONENTS Test #1

Solve/Simplify:

1. $3^3 =$ 2. $4^2 =$ 3. $5^3 =$

4. $8^{-1} =$ 5. $12^{-2} =$ 6. $5^{-3} =$

Rewrite as an exponential expression: (For example, $x \bullet x \bullet x = x^3$)

7. $y \bullet y =$ 8. $10 \bullet 10 \bullet 10 \bullet 10 =$ 9. $x \bullet x \bullet y \bullet y =$

10. $a \bullet a \bullet b =$ 11. $3 \bullet 3 \bullet 3 =$ 12. $c \bullet c \bullet c \bullet c \bullet c =$

Solve/Simplify:

13. $(-2)^2 =$ 14. $(-2)^{-2} =$ 15. $(-12)^3 =$

16. $3 + (4)^3 =$ 17. $-8 \bullet x^2 =$ 18. $15 - (-4)^3 =$

19. $5^3 - 4(35 \div 5)^2 =$ 20. $(-4)^3 + 5^2 - (3 \times 2)^2 =$

Write in scientific notation:

21. 45,000 = 22. 675,200,000 =

23. 1,240,000 = 24. 600 =

25. 0.00234 = 26. 0.000005=

27. 0.04589 = 28. 0.78 =

Solve/Simplify:

29. $2^4 \bullet 2^2 =$ 30. $3^2 \bullet 3^2 =$

31. $(x^4)^2 =$ 32. $(3^2)^4 =$

33. $(4^2)^{-3} =$ 34. $x^4/x^2 =$

35. $(2a^2)^3 =$ 36. $c^8/c^{12} =$

37. $(2xy^3z^5)^2 =$ 38. $(3x)^{-3} =$

39. $x^5/x^{-3} =$ 40. $(2y^4/3x^3)^{-2} =$

41. $\frac{20x^{-2}y^5}{5x^3y^3} =$ 42. $\frac{27a^{-3}bc^3}{9ab^4} =$ 43. $\frac{4x^7y^{-2}z^2}{12x^3yz^5} =$

BE AWESOME! Practice AS MUCH AS POSSIBLE! You deserve THE BEST GRADE POSSIBLE!

You = AMAZING

EXPONENTS Test #2

Solve/Simplify:

1. $4^3 =$ 2. $(-1)^2 =$ 3. $8^3 =$

4. $3^{-1} =$ 5. $5^{-2} =$ 6. $7^{-3} =$

Rewrite as an exponential expression: (For example, $x \bullet x \bullet x = x^3$)

7. $3 \bullet 3 =$ 8. $12 \bullet 12 \bullet 12 =$ 9. $a \bullet a \bullet c \bullet c =$

10. $4 \bullet 4 \bullet b \bullet b =$ 11. $x \bullet x \bullet y =$ 12. $x \bullet y \bullet x \bullet y \bullet x =$

Solve/Simplify:

13. $(-5)^2 =$ 14. $(-1)^{-2} =$ 15. $(-11)^3 =$

16. $7 + (-3)^3 =$ 17. $12 \bullet y^2 =$ 18. $18 + (-4)^3 =$

19. $3^3 - 7(45 \div 5)^2 =$ 20. $(-2)^3 + 4^2 - (3 \times 2)^3 =$

Write in scientific notation:

21. 7,000 = 22. 876,200,000 =

23. 18,000 = 24. 900 =

25. 0.00895 = 26. 0.000003=

27. 0.09728 = 28. 0.32 =

Solve/Simplify:

29. $2^4 \bullet 3^2 =$ 30. $4^2 \bullet 5^2 =$

31. $(y^{-3})^2 =$ 32. $(4^2)^{-2} =$

33. $(-6)^{-2} =$ 34. $a^6/a^8 =$

35. $(3x^5)^3 =$ 36. $b^5/b^2 =$

37. $(5x^4y^{-3}z^2)^2 =$ 38. $(5x^{-4})^{-2} =$

39. $y^7/y^{-2} =$ 40. $(5a^3/2c^3)^{-3} =$

41. $\frac{10x^7y^5}{5x^2y^6} =$ 42. $\frac{15a^{-2}b^2}{9ab^4} =$ 43. $\frac{9x^3y^{-7}z^2}{12x^3y^2z^5} =$

Step by step: See each problem in SUCCESS BOOK solved step-by-step in WEBINAR

Exponents Test #1 Answers:

1. 27
2. 16
3. 125
4. 1/8
5. 1/144
6. 1/125
7. y^2
8. 10^4
9. x^2y^2
10. a^2b
11. 3^3
12. c^5
13. 4
14. 1/4
15. -1728
16. 67
17. $-8x^2$
18. 79
19. -71
20. -75
21. 4.5×10^4
22. 6.752×10^8
23. 1.24×10^6
24. 6.9×10^2
25. 2.34×10^{-3}
26. 5.0×10^{-6}
27. 4.589×10^{-2}
28. 7.8×10^{-1}
29. 64
30. 81
31. x^8
32. 6561
33. 1/4096
34. x^2
35. $8a^6$
36. $1/c^4$
37. $4x^2y^6z^{10}$
38. $1/27x^3$
39. x^8
40. $9x^6/4y^8$
41. $4y^2/x^5$
42. $3c^3/a^4b^3$
43. $x^4/3y^3z^3$

Exponents Test #2 Answers:

1. 64
2. 1
3. 512
4. 1/3
5. 1/25
6. 1/343
7. 3^2
8. 12^3
9. a^2c^2
10. 4^2b^2
11. x^2y
12. x^3y^2
13. 25
14. 1
15. -1331
16. -20
17. $12y^2$
18. -46
19. -540
20. -208
21. 7.0×10^3
22. 8.762×10^8
23. 1.8×10^4
24. 9.0×10^2
25. 8.95×10^{-3}
26. 3.0×10^{-6}
27. 9.728×10^{-2}
28. 3.2×10^{-1}
29. 144
30. 400
31. $1/y^6$
32. 1/16
33. 1/36
34. $1/a^2$
35. $27x^{15}$
36. b^3
37. $25x^8z^4/y^6$
38. $x^8/25$
39. y^9
40. $8c^9/125a^9$
41. $2x^5/y$
42. $5/3a^3b^2$
43. $3/4y^9z^3$

Step by step: See each problem in SUCCESS BOOK solved step-by-step in WEBINAR

EQUATIONS Test #1

Solve for x:

1. $x + 5 = 11$
2. $x - 7 = 23$
3. $x + 9 = 14$
4. $8x + 8 = 7x - 10$
5. $9 + x = 17$
6. $2x + 6 = 3x - 11$
7. $4x = 24$
8. $20 = -10x$
9. $3x = -33$
10. $\frac{x}{9} = -6$
11. $\frac{x}{2} = 13$
12. $\frac{x}{-3} = 22$
13. $6x + 2 = 20$
14. $2x + 3 = 5x + 9$
15. $7x - 5 = 33$
16. $4(x + 1) = 18$
17. $3(5x - 2) = 24$
18. $-2(2x + 3) = 5$
19. $7x + 4(3x - 10) = 11x - 5$
20. $-(1 + 7x) - 6(-7 - x) = 36$
21. $|x| = 2$
22. $|3x| = 21$
23. $4|x - 5| = 17 - (-7)$
24. $8 + 2|2x + 3| = 22$

BE AWESOME! Practice AS MUCH AS POSSIBLE! You deserve THE BEST GRADE POSSIBLE!

EQUATIONS Test #2

Solve for x:

1. $x + 8 = 14$
2. $x - 2 = 17$
3. $x + 4 = -5$
4. $12x + 3 = 11x - 8$
5. $-5 + x = 9$
6. $4x + 2 = 5x - 16$
7. $-6x = 36$
8. $-120 = -20x$
9. $5x = -45$
10. $x/8 = 7$
11. $x/2 = -50$
12. $x/-5 = -12$
13. $-3x + 2 = 23$
14. $-x - 5 = 7x + 11$
15. $2x - 3 = -31$
16. $-5(x - 1) = 35$
17. $2(-3x - 7) = 24$
18. $-6(x + 2) = 18$
19. $3x - 2(2x + 7) = 9x - 9$
20. $(9 - 6x) + 2(3 + 2x) = 5x - 21$
21. $|x| = 9$
22. $|7x| = 49$
23. $6|x + 3| = 20 - (-4)$
24. $8 + 5|4x - 1| = 23$

Step by step: See each problem in SUCCESS BOOK solved step-by-step in WEBINAR

Equations Test #1 Answers:

1. x = 6
2. x = 30
3. x = 5
4. x = -18
5. x = 8
6. x = 17
7. x = 6
8. x = -2
9. x = -11
10. x = -54
11. x = 26
12. x = -66
13. x = 3
14. x = -2
15. x = 38/7 or 5 3/7
16. x = 7/2 or 3 1/2
17. x = 2
18. x = -2 3/4
19. x = 4 3/8
20. x = 5
21. x = 2, x = -2
22. x = 7, x = -7
23. x = 11, x = -1
24. x = 2, x = -5

Equations Test #2 Answers:

1. x = 6
2. x = 19
3. x = -9
4. x = -11
5. x = 14
6. x = 18
7. x = -6
8. x = 6
9. x = -9
10. x = 56
11. x = -100
12. x = 60
13. x = -7
14. x = -2
15. x = -14
16. x = 8
17. x = -19/3 or -6 1/3
18. x = -5
19. x = -1/2
20. x = 5 1/7
21. x = 9, x = -9
22. x = 7, x = -7
23. x = 1, x = -7
24. x = 1, x = -1/2

GRAPHING Test #1

Graph the following linear equations using A TABLE:

1. $y = 2x + 1$
2. $y = \frac{4}{5}x - 2$
3. $x + y = 3$

Graph the following linear equations using SLOPE INTERCEPT FORM (y = mx + b):

4. $y = \frac{3}{4}x - 5$
5. $y = x + 4$
6. $y = \frac{-2}{3}x + 1$

Find the x and y intercepts of each linear equation, then graph:

7. $4x - 2y = 12$
8. $-3x + 6y = 18$
9. $5x + 2y = -10$

Find the slope of the line that passes through two given points:

10. (3, -2) (5, 4)
11. (-1, -6) (4, -6)
12. (-3, -1) (4, 5)

Rewrite the following linear equations in slope intercept form (y = mx + b):

13. $2x - 3y = 6$
14. $-x + 5y = 15$
15. $3y - 5x = -2x + 8$

Write the equation of the line containing the given point and slope in POINT-SLOPE FORM:
$(y - y_1 = m(x - x_1)$

16. (3, -8), m= -2
17. (-5, 0), m= $\frac{2}{3}$
18. (-1, 4), m= $\frac{1}{5}$

Write the equation of the line containing the given points in POINT-SLOPE FORM:

19. (-9, 4) (-6, -3)
20. (0, -2) (4, 5)
21. (7, -1) (2, 3)

Write the equation of the line that passes through the given point and is a) parallel and b) perpendicular to the given line in POINT-SLOPE FORM:

22. (4, -2)
    $y = -3x + 7$
23. (-1, -6)
    $2x - 3y = 12$
24. (5, -5)
    $-5x + 2y = 7$

BE AWESOME! Practice AS MUCH AS POSSIBLE! You deserve THE BEST GRADE POSSIBLE!

You = AMAZING

GRAPHING Test #2

Graph the following linear equations using A TABLE:

1. $y = {}^{-}\frac{1}{2}x - 5$
2. $y = 4x - 2$
3. $3x + 4y = 12$

Graph the following linear equations using SLOPE INTERCEPT FORM (y = mx + b):

4. $y = \frac{1}{4}x - 1$
5. $y = -3x + 3$
6. $y = {}^{-}\frac{3}{5}x + 6$

Find the x and y intercepts of each linear equation, then graph:

7. $3x - 6y = -12$
8. $-6x + 5y = 30$
9. $x - 2y = 4$

Find the slope of the line that passes through two given points:

10. (2, -3) (1, 5)
11. (-3, -5) (2, -4)
12. (1, -1) (0, 7)

Rewrite the following linear equations in slope intercept form (y = mx + b):

13. $-4x - 3y = 9$
14. $2x + 4y = 12$
15. $y - 8x = -x + 12$

Write the equation of the line containing the given point, slope in POINT-SLOPE FORM:
$(y - y_1 = m(x - x_1))$

16. (2, -5), $m = -1$
17. (3, 1), $m = {}^{-}\frac{1}{3}$
18. (4, -6), $m = \frac{3}{5}$

Write the equation of the line containing the given points in POINT-SLOPE FORM:

19. (-8, 2) (-6, -6)
20. (-3, 0) (-5, 4)
21. (3, 7) (7, 3)

Write the equation of the line that passes through the given point and is a) parallel and b) perpendicular to the given line in POINT-SLOPE FORM:

22. (3, -1)
    $y = {}^{-}\frac{1}{4}x + 7$
23. (2, -5)
    $-x - 2y = 7$
24. (0, 4)
    $3x + 2y = -8$

Step by step: See each problem in SUCCESS BOOK solved step-by-step in WEBINAR

Algebra is EASY and YOU CAN DO IT

Graphing Test #1 Answers:

7. x = 3, y = -6

8. x = -6, y = 3

9. x = -2, y = -5

10. m = 3

11. m = 0

12. m = 6/7

13. y = 2/3 x - 2

14. y = 1/5 x + 3

15. y = x + 3

16. y + 8 = -2(x - 3)

17. y - 0 = 2/3(x + 5)

18. y - 4 = 1/5(x + 1)

19. y + 3 = -7/3(x + 6)

20. y + 2 = 7/4(x - 0)

21. y - 3 = -4/5(x - 2)

22. a) y + 2 = -3(x - 4)

b) y + 2 = 1/3(x - 4)

23. a) y + 6 = 2/3(x + 1)

b) y + 6 = -3/2(x + 1)

24. a) y + 5 = 5/2(x - 5)

b) y + 5 = -2/5(x - 5)

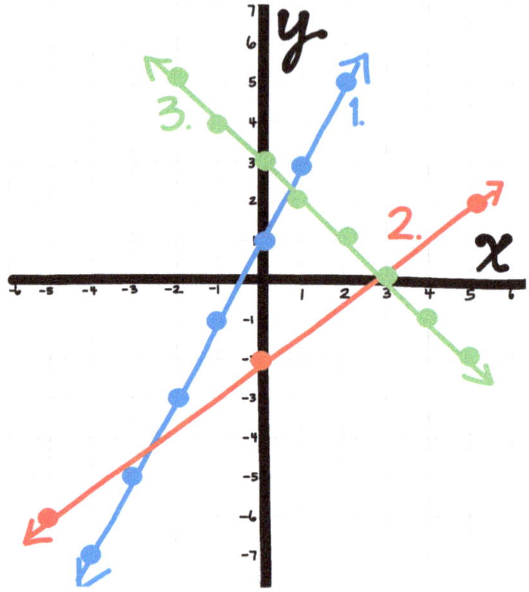

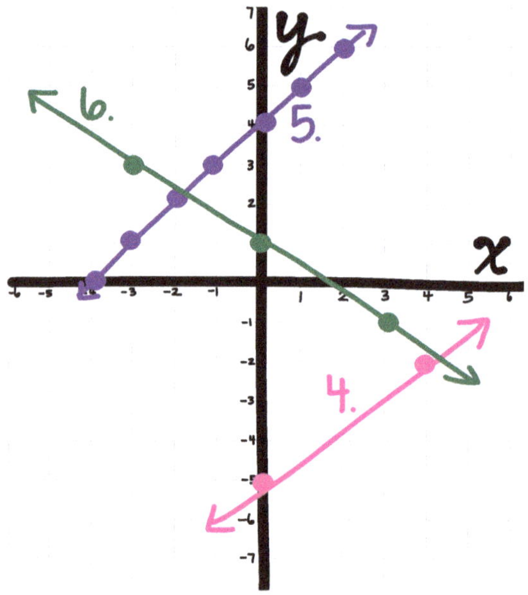

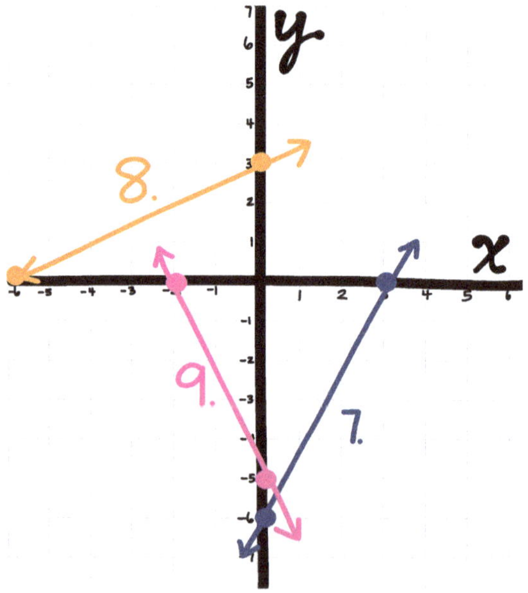

BE AWESOME! Practice AS MUCH AS POSSIBLE! You deserve THE BEST GRADE POSSIBLE!

You = AMAZING

Graphing Test #2 Answers:

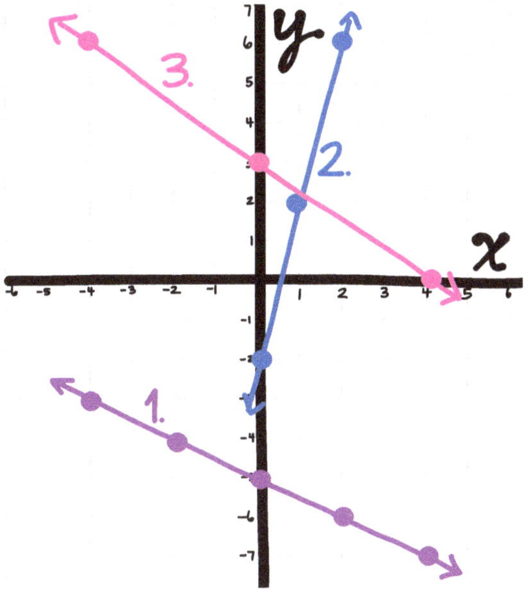

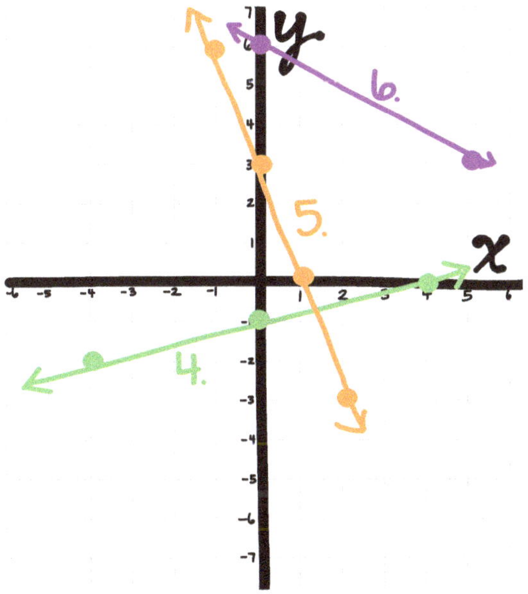

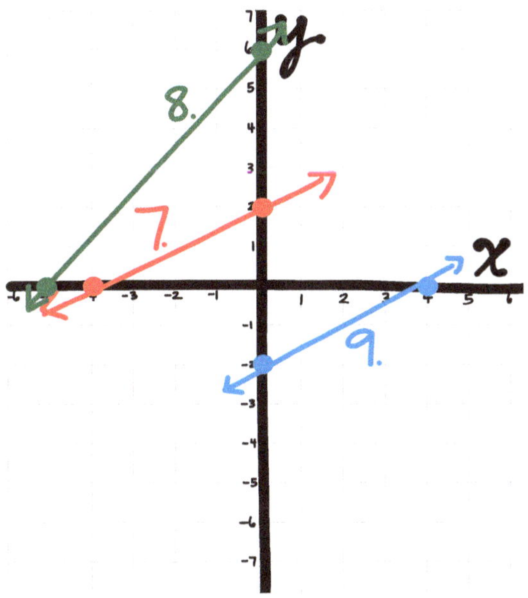

7. x = -4, y = 2

8. x = -5, y = 6

9. x = 4, y = -2

10. m = -8

11. m = -1

12. m = -8

13. y = -4/3 x - 3

14. y = -1/2 x + 3

15. y = 7x + 12

16. y + 5 = -1(x - 2)

17. y - 1 = -1/3(x - 3)

18. y + 6 = 3/5(x - 4)

19. y -2 = -4(x + 8)

20. y - 0 = -2(x + 3)

21. y - 7 = -1(x - 3)

22. a) y + 1 = -1/4(x - 3)

b) y + 1 = 4(x - 3)

23. a) y + 5 = -1/2(x - 2)

b) y + 5 = 2(x - 2)

24. a) y - 4 = -3/2(x - 0)

b) y - 4 = 2/3(x - 0)

INEQUALITIES Test #1

Solve the following inequalities:

1. $x + 5 < -4$
2. $x - 7 \ge -3$
3. $x + 9 > 6$
4. $8x + 8 \le 4x - 10$
5. $3 - 4x > 8x + 7$
6. $-12x + 8 < 20x$
7. $4x \ge 24$
8. $20 < -10x$
9. $3x > -33$
10. $\frac{x}{9} \le -6$
11. $\frac{x}{2} > 13$
12. $\frac{2x}{-3} \le 22$
13. $6(x + 1) < 2$
14. $2(-3x + 3) \ge 5x + 9$
15. $7x - 5 < 33$
16. $-3 < x + 2 < 4$
17. $-5 \le 2x - 4 \le 20$
18. $-3 < -3x \le 9$
19. $-52 < 4(3x - 10) < 11 - 3$
20. $-6 < -3(2x + 4) \le 27 + (-3)$
21. $|x| < 2$
22. $|3x| \ge 21$
23. $4|x - 5| > 17 - (-7)$

Graph the following inequalities on a number line:

24. $x > -4$
25. $x < 5$
26. $-3 < x \le 4$
27. $2 \le x \le 3$

Graph the following inequalities on a grid:

28. $y > 3$ and $x < -1$
29. $y \le \frac{1}{2}x - 3$
30. $-4x - 2y < 12$

BE AWESOME! Practice AS MUCH AS POSSIBLE! You deserve THE BEST GRADE POSSIBLE!

INEQUALITIES Test #2

Solve the following inequalities:

1. $x - 2 < -7$
2. $x + 1 \geq -8$
3. $x + 11 > 5$
4. $12x + 3 \leq -3x - 12$
5. $4 - 4x > 7x + 7$
6. $-2x + 8 < -26x$
7. $3x \geq -27$
8. $21 < -7x$
9. $2x > -18$
10. $\frac{3x}{4} \leq -5$
11. $\frac{x}{2} > -2$
12. $\frac{4x}{5} \leq 12$
13. $-4(x - 3) < 11$
14. $3(-2x + 1) \geq x + 7$
15. $\frac{1}{2}x - 4 < 12$
16. $-1 < x - 3 < 5$
17. $-4 \leq 4x - 2 \leq 18$
18. $-2 < -6x \leq 10$
19. $3 < 4(2x + 3) < 25$
20. $-5 < 4(-x + 7) \leq 20$
21. $|x| < 5$
22. $|4x| \geq 16$
23. $3|x + 1| > 39$

Graph the following inequalities on a number line:

24. $x > -1$
25. $x < 3$
26. $-2 < x \leq -1$
27. $8 \leq x \leq 19$

Graph the following inequalities on a grid:

28. $y > -2$ and $x < -2$
29. $y \leq \frac{2}{3}x + 5$
30. $3x - 2y < 14$

Inequalities Test #1 Answers:

1. $x < -9$
2. $x \geq 4$
3. $x > -3$
4. $x \leq -4\ 1/2$
5. $x < -1/3$
6. $x > 1/4$
7. $x \geq 6$
8. $x < -2$
9. $x > -11$
10. $x \leq -54$
11. $x > 26$
12. $x \geq -33$
13. $x < -2/3$
14. $x \leq -3/11$
15. $x < 5\ 3/7$
16. $-5 < x < 2$
17. $-1/2 \leq x \leq 12$
18. $-3 \leq x < 1$
19. $-1 < x < 4$
20. $-6 \leq x < -1$
21. $x < 2, x > -2$
22. $x \geq 7, x \leq -7$
23. $x > 11, x < -1$

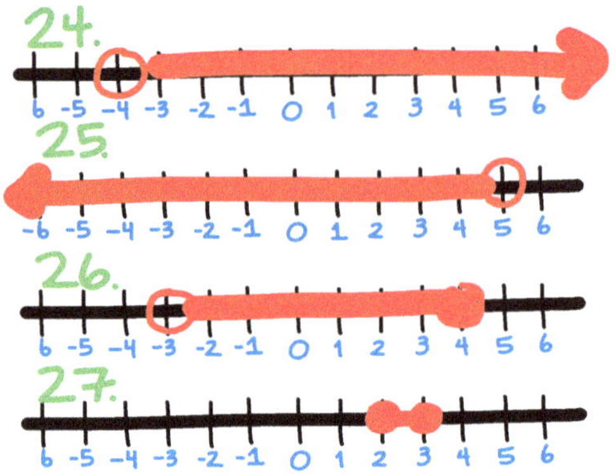

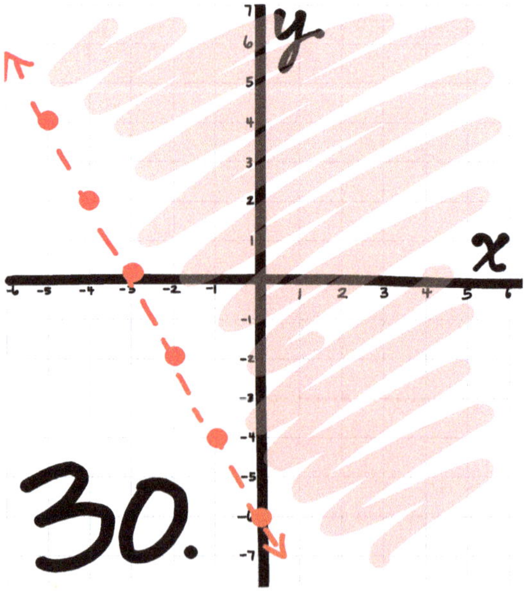

BE AWESOME! Practice AS MUCH AS POSSIBLE! You deserve THE BEST GRADE POSSIBLE!

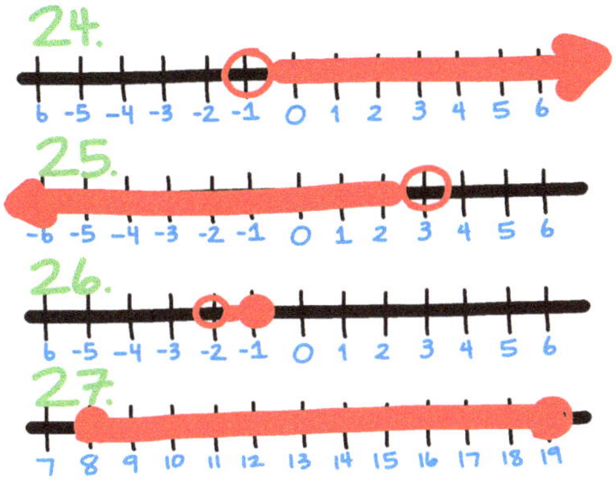

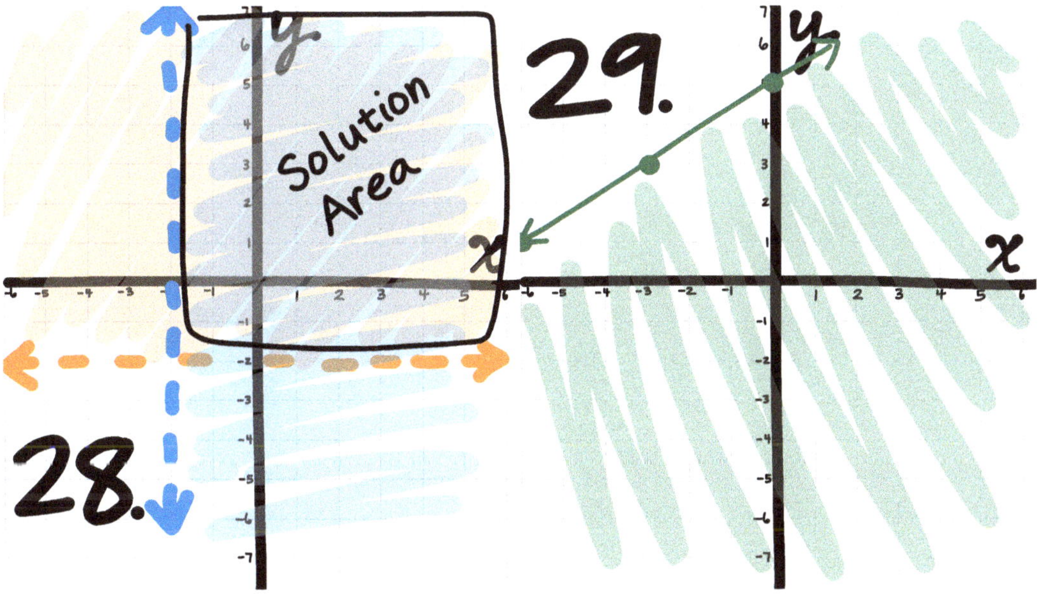

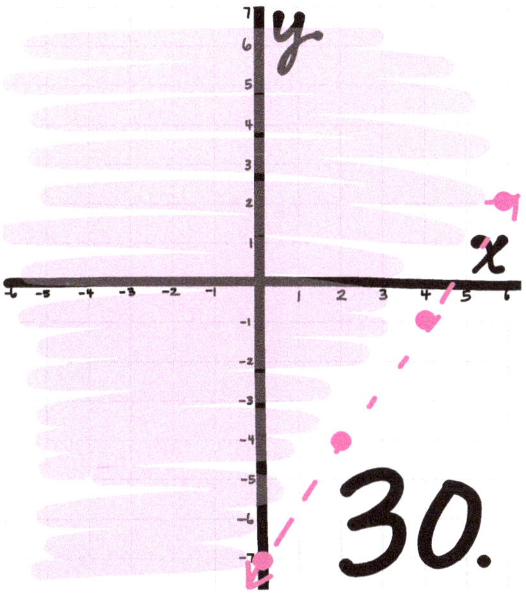

Inequalities Test #2 Answers:

1. x < -5
2. x ≥ -9
3. x > -6
4. x ≤ -1
5. x < -3/11
6. x < -1/3
7. x ≥ -9
8. x < -3
9. x > -9
10. x ≤ -20/3 or -6 2/3
11. x > -4
12. x ≤ 15
13. x > 1/4
14. x ≤ -4/7
15. x < 32
16. 2 < x < 8
17. -1/2 ≤ x ≤ 5
18. -5/3 ≤ x < 1/3
19. -9/8 < x < 13/8
20. 2 ≤ x < 33/4
21. x < 5, x > -5
22. x ≥ 4, x ≤ -4
23. x > 12, x < -14

www.ingramcontent.com/pod-product-compliance
Lightning Source LLC
Chambersburg PA
CBHW041433040426
42451CB00023B/3497

* 9 7 8 1 9 4 1 7 7 5 2 6 4 *